# Simula SpringerBriefs on Computing

Reports on Computational Physiology

Volume 14

**Editor-in-Chief**

Joakim Sundnes, Simula Research Laboratory, Oslo, Norway

**Series Editors**

Kimberly J. McCabe, Simula Research Laboratory, Oslo, Norway

Andrew D. McCulloch, Institute for Engineering in Medicine, University of California San Diego, La Jolla, California, USA; Department of Bioengineering, University of California San Diego, La Jolla, California, USA

Aslak Tveito, Simula Research Laboratory, Oslo, Norway

**Managing Editor**

Jennifer Hazen, Simula Research Laboratory, Oslo, Norway

## About this Series

In 2016, Springer and Simula launched an Open Access series called the Simula SpringerBriefs on Computing. This series aims to provide concise introductions to the research areas in which Simula specializes: scientific computing, software engineering, communication systems, machine learning and cybersecurity. These books are written for graduate students, researchers, professionals and others who are keenly interested in the science of computing, and each volume presents a compact, state-of-the-art disciplinary overview and raises essential critical questions in the field.

Simula's focus on computational physiology has grown considerably over the last decade, with the development of multi-scale mathematical models of excitable tissues (brain and heart) that are becoming increasingly complex and accurate. This subseries represents a new branch of the SimulaSpringer Briefs that is specifically focused on computational physiology. Each volume in the series will introduce and explore one or more sub-fields of computational physiology, and present models and tools developed to address the fundamental research questions of the field. Whenever possible, the software used will be made publicly available.

By publishing the Simula SpringerBriefs on Computing and the sub-series on computational physiology, Simula Research Laboratory acts on its mandate of emphasizing research education. Books in the series are published by invitation from one of the editors, and authors interested in publishing in the series are encouraged to contact any member of the editorial board.

Karoline Horgmo Jæger • Aslak Tveito

# Differential Equations for Studies in Computational Electrophysiology

simula

Karoline Horgmo Jæger
Simula Research Laboratory
Oslo, Norway

Aslak Tveito
Simula Research Laboratory
Oslo, Norway

ISSN 2512-1677          ISSN 2512-1685  (electronic)
Simula SpringerBriefs on Computing
ISSN 2730-7735          ISSN 2730-7743  (electronic)
Reports on Computational Physiology
ISBN 978-3-031-30851-2          ISBN 978-3-031-30852-9  (eBook)
https://doi.org/10.1007/978-3-031-30852-9

Mathematics Subject Classification (2020): 92-01, 92C05, 34-01, 35-01, 65-M99

This Springer imprint is published by the registered company Springer Nature Switzerland AG
The registered company address is: Gewerbestrasse 11, 6330 Cham, Switzerland

# Series Foreword

Dear reader,

the series *Simula SpringerBriefs on Computing* was established in 2016, with the aim of publishing compact introductions and state-of-the-art overviews of select fields in computing. Research is increasingly interdisciplinary, and students and experienced researchers both often face the need to learn the foundations, tools, and methods of a new field. This process can be demanding, and typically involves extensive reading of multidisciplinary publications with different notations, terminologies and styles of presentation. The briefs in this series are meant to ease the process by explaining important concepts and theories in a specific interdisciplinary field without assuming extensive disciplinary knowledge and by outlining open research challenges and posing critical questions in the field.

Simula has a major research program in computational physiology that includes a long and close collaboration with the University of California (UC) San Diego. To reflect this research focus, we established in 2020 a new subseries entitled *Simula Springer Briefs on Computing - Reports on Computational Physiology*. The subseries includes both introductory and advanced texts on select fields of computational physiology, designed to advance interdisciplinary scientific literacy and promote effective communication and collaboration in the field. This subseries is also the outlet for collections of reports from the annual Summer School in Computational Physiology, organized by Simula, University of Oslo, and UC San Diego. The school starts in June each year with students spending two weeks in Oslo learning the principles underlying mathematical models commonly used in studying the heart and the brain. During their stay in Oslo, students are assigned a research project to work on over the summer. In August, they travel to San Diego for another week of training and project work, and a final presentation of their findings. Every year, we have been impressed by the students' creativity and we often see results that could lead to a scientific publication. Starting with the 2021 edition of the summer school, we have taken the course one step further by having each team conclude their project with a scientific report that can pass rigorous peer review as a publication in this subseries.

All items in the main series and the subseries are published within the SpringerOpen framework, as this will allow authors to use the series to publish an initial version of their manuscript that could subsequently evolve into a full-scale book on a broader theme. Since the briefs are freely available online, the authors do not receive any direct income from the sales; however, remuneration is provided for every completed manuscript. Briefs are written on the basis of an invitation from a member of the editorial board.

Suggestions for possible topics are most welcome, and interested authors are encouraged to contact a member of the editorial board.

March 2023

*Dr. Joakim Sundnes*
sundnes@simula.no

*Dr. Kimberly J. McCabe*
kimberly@simula.no

*Dr. Andrew McCulloch*
amcculloch@ucsd.edu

*Dr. Aslak Tveito*
aslak@simula.no

**Series Editor for this Volume**

Joakim Sundnes, Simula Research Laboratory, Oslo, Norway

# Preface

## Why would you want to read these notes?

When something is very large, very small or very complex, it's often helpful to represent it using a model in order to understand what is going on. Nature offers many examples of situations where direct interrogations are difficult, and a model can provide insight into the phenomena of interest. Physiology, for example, is a field that is complex and sometimes poorly understood, and where mathematical models are frequently used to increase understanding. These models are often formulated in terms of differential equations. However, many students who want to learn state-of-the-art physiology may not be familiar with differential equations. They are probably able to use software tools that solve these equations in numerical simulations, but they may not fully understand the underlying principles. Our goal with these notes is to provide a simple introduction to differential equations and give examples of how to solve them. Differential equations is a vast area of active research, and we can only provide a glimpse. But, we hope that these notes will provide a basic understanding of the principles, such that computational codes will appear less like a black box, and more like something that is comprehensible.

## The Simula Summer School in Computational Physiology

Every year Simula Research Laboratory organizes a summer school in Computational Physiology together with the University of California, San Diego (UCSD) and the University of Oslo (UiO). The students come to Oslo in June and learn about models and software used in computational physiology. Next, the students are divided into groups and work on different projects through the summer. All groups are assigned one or more mentors who help the students with their investigations. In August, all students and mentors meet again at UCSD and continue their work on the project. In addition, they attend guest lectures at UCSD and a two-day workshop on scientific writing organized by experienced editors of *Nature*.

The models and software used in the course are state-of-the art and the students complete the course by writing a scientific report that is published in a Simula SpringerBriefs on Computing in the sub-series on Computational Physiology. In order to reach the advanced level of using state-of-the-art methods and software, there is not enough time to cover the details of all the subjects covered in the course. The students are most often enrolled in MSc., PhD or Post Doc programs in universities around the world. Their scientific backgrounds vary greatly. There are students from theoretical disciplines like computer science, scientific computing, statistics or mathematics, and from more classical science disciplines like physics, biology, or medicine. And perhaps the largest group of students comes with a background in bioengineering. We notice that the students easily follow 'their' part of the course and struggle with the parts they are unfamiliar with.

Scientific work in computational physiology is inherently interdisciplinary so meeting this reality in the summer school is proper training. Typical research projects comprise elements of many disciplines and thus it is very common to not fully understand all elements of a project, and almost no project is completed by only one person. Similarly, it is exceptionally rare to see a single-authored paper in computational physiology, or in computational science in general for that matter. So, learning to communicate across disciplines is very useful, but also very challenging.

The most common mathematical machinery used to model physiology (and physics in general) is differential equations. Most of the models introduced in the summer school is founded, in one way or another, on differential equations. Elements of this subject are taught at every university in the world and students from mathematics, physics or scientific computing most likely have a course in this subject. But students from biology, medicine or even computer science rarely know much about differential equations. Now, the summer school is very streamlined and the software is usually prepared and can be used without a complete understanding of the underlying models, but it is clearly advantageous to know, at least intuitively, the foundation of the models and solution methods that are used in the course.

## How are the notes organized?

We have organized these notes in two parts. In the first part (Chapters 1–7), we introduce the concept of differential equations and numerical methods. To keep the exposition as simple as possible, we use simple, unitless equations as examples in these chapters. In the second part (Chapters 8–12), we apply the techniques introduced in the first part of the book to a selection of models of electrophysiology.

More specifically, we start these notes by considering a very simple differential equation. For this equation, we introduce a numerical method and we study the error of the method. The simplicity of the equation allows us to study these important concepts in a very explicit manner. Next, we move on to systems of ordinary differential equations and we show how the FitzHugh-Nagumo model can be solved numerically. Partial differential equations are then introduced and we show how the diffusion equation can be solved numerically. By combining the diffusion equation and the FitzHugh-Nagumo model, we introduce traveling wave solutions and show how a reaction-diffusion system can be handled numerically.

At the beginning of the second part of the notes, in Chapter 8, we introduce the Hodgkin-Huxley equations and from this chapter we start using units for all quantities involved. The Hodgkin-Huxley equations are the most famous system of equations in physiology. The equations model the action potential of an axon and have been proved to represent that process with great accuracy. We also consider a similar model for a cardiac action potential. The membrane models of cardiac electrophysiology have evolved into a very complex matter, but the structure remains very similar to the Hodgkin-Huxley model.

After being familiarized with the Hodgkin-Huxley membrane model, we combine it with the cable equation to model the propagation of an action potential along an axon. Next, we introduce the bidomain and the monodomain models in two spatial dimensions. The monodomain equation is very similar to the cable equation, but the numerical solution will be a bit more complicated because we consider a 2D problem. The bidomain model is more complex than the monodomain model, but we will see that by using operator splitting, solution methods become quite straightforward. Never heard of operator splitting? You will learn about it from these notes. It is a very powerful technique used to break complicated problems into problems we already know how to deal with. We will explain it in Chapter 7 and show some applications later in the notes.

One major issue with the monodomain and bidomain models is that they represent averaged quantities in the sense that the cardiomyocytes are not present in the model. In fact, both the extracellular (E) space, the cell membrane (M) and the intracellular (I) space are assumed to exist everywhere. That is a bold assumption, so we will also introduce the EMI model where all these elements (E, M, and I) are explicitly part of the model.

The monodomain and bidomain models provide reasonable approximations of cardiac electrophysiology at the mm-scale, and the EMI model addresses electrophysiology at the $\mu$m-scale. The next level is the nm scale[1]. The relevant equations at the nm-scale are the Poisson-Nernst-Planck (PNP) equations. From a computational perspective, the PNP equations are extremely challenging and can, at present, only be used to study very small regions; one complete cardiomyocyte is far too large to be modeled by the PNP equations. We will show how to solve these equations in the final chapter using operator splitting and finite differences.

In order to keep these notes relatively brief, many details about the methods and models introduced must inevitably be left out. At the end of most chapters, we therefore include a section called 'Comments and Further Reading', providing some comments on these details and suggestions for further reading.

---

[1] Note here that meter is denoted by m, and mm means millimeter or $10^{-3}$m, $\mu$m is micrometer and means $10^{-6}$m, and, finally, nm means nanometer or $10^{-9}$m. If you ever need help with units, we recommend www.wolframalpha.com.

# Why do we use the finite difference method?

There are many ways to solve differential equations. In computational physiology, the finite element method may very well be the most popular alternative. The reason for this is that the geometry of the computational domain (e.g., the heart) is quite complex and thus very difficult to represent using a finite difference method. Finite difference methods are easiest to work with when the computational mesh is *very regular*, but the finite element method is constructed to allow for highly irregular meshes. Both finite element methods and finite volume methods are successfully applied to simulate complex phenomenas on complex geometries. The finite difference method is successfully applied to simulate complex dynamics but the code quickly becomes clunky in the presence of geometries that are any more complex than a cuboid. Nevertheless, we have chosen to focus entirely on the finite difference method in these notes and the reason for this is simplicity. The method is more or less completely defined by simply replacing derivatives by differences. Therefore, we stick to simple geometries and use finite differences. In the summer school, much more advanced simulations, using finite elements, will be performed, but we still think it is useful to know how this in principle can be done using the simplest possible method.

# It's open access

These notes are printed in Simula SpringerBriefs on Computing in the sub-series on Computational Physiology. That means that the notes can be downloaded for free. The software (Matlab) used to generate all figures and tables in these notes are freely available online. The codes are written primarily for clarity and less emphasis is put on efficiency. If you find a bug in the codes or an error in these notes, please send a mail to aslak@simula.no. The software associated with these notes can be found at: https://github.com/karolihj/differential-equations-book-2023

# Acknowledgement

The authors extend their gratitude to Joakim Sundnes, Kimberly McCabe, Lars Tveito, and Jennifer Hazen for their invaluable comments to the manuscript. They also acknowledge the financial support provided by Simula Research Laboratory and the SUURPh program funded by the Norwegian Ministry of Education and Research. In addition, it is a pleasure to express our appreciation of the long-term collaboration with Dr. Martin Peters at SpringerNature.

Oslo,                                                      *Karoline Horgmo Jæger*
February 2023                                                      *Aslak Tveito*

# Contents

# Part I
# Tools for Differential Equations and Numerical Methods

# Chapter 1
# Getting Started

In this introductory chapter we will introduce two essential concepts: differential equations and numerical methods. If you want to spend as little energy as possible (don't be ashamed of that - energy preservation is both fashionable and a fundamental property of many biological mechanisms - it's fine) you can get a good overview from this chapter alone.

## 1.1 What Is a Differential Equation?

You have no doubt seen an *algebraic* equation. A typical algebraic equation may look like

$$x^2 - 4x + 3 = 0 \tag{1.1}$$

with solutions $x = 1$ and $x = 3$. So, solutions of algebraic equations are numbers. The solutions of differential equations, on the other hand, are functions. One very simple differential equation is given by

$$y'(t) = y(t). \tag{1.2}$$

Here, we typically assume that $t$ represents time, and that the equation (1.2) describes how the function $y$ changes with time. Suppose we also know that the solution is 1 at $t = 0$, that is

$$y(0) = 1. \tag{1.3}$$

Such a condition is generally referred to as an *initial condition* and is needed in order to find a unique solution of the problem. In this case, the solution is given by the function

$$y(t) = e^t. \tag{1.4}$$

It is straightforward to verify this. We simply note that $y(0) = e^0 = 1$ and, by differentiation, that $y'(t) = e^t = y(t)$. So all is good; we have found the solution of our first differential equation.

© The Author(s) 2023
K. Horgmo Jæger, A. Tveito, *Differential Equations for Studies in Computational Electrophysiology*,
Simula SpringerBriefs on Computing 14, https://doi.org/10.1007/978-3-031-30852-9_1

For a long time, differential equations were solved in this way. Formulas were derived using pencil and paper. If the equations were very complex, they were simplified in order to be solved by a formula and people spent whole academic careers deriving such approximate solutions of differential equations. The reason for this was that the solutions of the equations could bring critical insight into a phenomenon modeled by the equation. Nowadays, differential equations are solved by computers. In some cases analytical formulas can be derived and then it is usually done by computing systems like Mathematica or Maple, or other tools for symbolic computations. But the more common approach is to solve the equations numerically. In order to do that, we need to transform the equations into a form that is suitable for computers. Almost all differential equations that are solved in computational physiology are solved by computers. It is the main purpose of these notes to teach you the basics of how that is done, and we might as well start with the very simple case that we've already introduced.

## 1.2 What Is a Numerical Method?

*A numerical method is a way to solve a mathematical problem on a computer.* We will see that one way to prepare differential equations for solution on computers is to replace derivatives by differences. That may not come as a surprise since you may remember from calculus that the derivative was introduced as the limit of a difference. Concretely, for a smooth[1] function $f = f(t)$, the derivative is, per definition,

$$f'(t) = \lim_{\Delta t \to 0} \frac{f(t + \Delta t) - f(t)}{\Delta t}. \tag{1.5}$$

Instead of going all the way to zero, we can settle for a small value of $\Delta t$ and use the approximation

$$f'(t) \approx \frac{f(t + \Delta t) - f(t)}{\Delta t}. \tag{1.6}$$

Here, $t$ is time, and we are interested in solving the equation from $t = 0$ (where the solution, $y$, is known to be 1) until some $t = T$, for example $T = 1$. For some fixed value of $\Delta t$, it is useful to define a number of discrete points in time,

$$t_n = n \times \Delta t, \tag{1.7}$$

for $n = 0, ..., N$, where $N = T/\Delta t$. We note here that the step in time from $t_{n-1}$ to $t_n$ is $\Delta t$, and that the final time is given by $t_N = N \times \Delta t = T$. We want to compute a numerical approximation to the solution of the problem

---

[1] A lot could be said about *smooth functions*. Generally, regularity of functions is important in solving differential equations. But for these notes it is sufficient to simply think of smooth functions as – smooth.

$$y'(t) = y(t), \tag{1.8}$$
$$y(0) = 1, \tag{1.9}$$

at the time steps $\{t_n\}_{n=0}^N$ by replacing the derivative of $y$ by the formula given in (1.6). This replacement can be written as

$$\frac{y_{n+1} - y_n}{\Delta t} = y_n, \tag{1.10}$$
$$y_0 = 1, \tag{1.11}$$

where, in general, $y_n$ denotes an approximation of the solution of (1.8) at time $t_n$; i.e., $y_n \approx y(t_n)$. By rearranging (1.10), we find that

$$y_{n+1} = (1 + \Delta t)y_n, \tag{1.12}$$

and we refer to this as the *computational form* of the numerical scheme. Since $y_0 = 1$, we find $y_1 = 1 + \Delta t$, and $y_2 = (1 + \Delta t)y_1 = (1 + \Delta t)^2$, and so forth. In general, we have

$$y_n = (1 + \Delta t)^n. \tag{1.13}$$

So in this unusually simple case, we don't need a program to compute the numerical solution. That is very unusual and we use this example just to introduce the concepts. More commonly, we first need to compute the solution $y_1$ from the initial condition, $y_0$, using a formula like (1.12), and then insert this solution $y_1$ into (1.12) to find the solution $y_2$, and so on until we find $y_N$ using the previously computed $y_{N-1}$. This type of repetitive computation is usually most conveniently performed using a computer program.

In Fig. 1.1 we have plotted the analytical (i.e. exact) solution, $y$, together with the numerical solution, $y_n$, for some different choices of the time step, $\Delta t$. We observe that when $\Delta t$ is small, the numerical and analytical solutions are indistinguishable, but that for larger values of $\Delta t$, for example $\Delta t = 0.2$, there is a visible difference between the two solutions. The analytical solution is exact, which means that the difference between the two solutions are due to an error of the numerical method. In the next section, we will consider this error in some more detail.

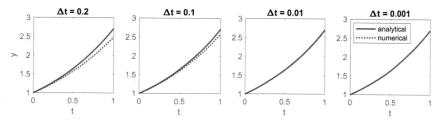

**Fig. 1.1** Analytical (solid line) and numerical (dotted line) solutions of the differential equation (1.8)–(1.9) for $t$ between 0 and 1 for some different values of $\Delta t$.

## 1.3 What Is the Error of a Numerical Method?

*The error of a numerical method is the difference between the numerical and the exact solution of the problem.* Since we are in the fortunate position of knowing both the analytical (exact) solution, $y(t)$, and the numerical solution, $y_n$, we can also compute the error[2] of the numerical solution defined by

$$E_n = |y(t_n) - y_n|. \qquad (1.14)$$

As observed in Fig. 1.1, the error depends on the size of the time step, $\Delta t$, which is natural since we are supposed to pass to the limit of $\Delta t = 0$ in the definition of the derivative (see (1.6)). When we approximate 0 by something small, we must be prepared for it to come with a price – and the price is the error we will encounter. To illustrate this, we put $T = 1$ and compare the analytical solution $y(T) = e$ with the numerical solution at $T = 1$ for several values of $\Delta t$. Since the final time is $T = 1$, we must have $N \times \Delta t = 1$ and therefore we compare

$$y_N = \left(1 + \frac{1}{N}\right)^N \qquad (1.15)$$

with the exact solution $y(1) = e$. In Table 1.1, we show the error $E_N = |y(1) - y_N|$ for $N = 5, 10, 100$, and $1000$. We also show $E_N/\Delta t$ and we note that this value seems to be more or less constant. It is thus evident that when $\Delta t$ goes to zero ($N$ goes to infinity), the numerical solution converges towards the analytical solution[3].

**Table 1.1** Error of the numerical solution of the differential equation (1.2)–(1.4) at $t = T = 1$ for different values of $\Delta t = \frac{T}{N}$. The error is defined as $E_N = |y(1) - y_N|$, where $y(1) = e$ is the analytical solution, and $y_N$ is the numerical solution, defined by (1.15).

| $N$ | $\Delta t$ | $E_N$ | $E_N/\Delta t$ |
|-----|-----------|-------|----------------|
| 5 | 0.2 | 0.23 | 1.15 |
| 10 | 0.1 | 0.125 | 1.25 |
| 100 | 0.01 | 0.0135 | 1.35 |
| 1000 | 0.001 | 0.00136 | 1.36 |

---

[2] The error defined here is referred to as the *absolute error*. An alternative is to consider the relative error given by

$$\frac{|y(t_n) - y_n|}{|y(t_n)|}.$$

[3] If this feels familiar, it may be because you learned in calculus that

$$\lim_{\varepsilon \to 0} (1 + \varepsilon)^{\frac{1}{\varepsilon}} = e.$$

We have learned that by replacing the derivative $y'(t)$ by a finite difference approximation $(y_{n+1} - y_n)/\Delta t$ we can find an approximate solution. Usually, the finite difference scheme must be implemented on a computer but in this very simple case, we can find a formula for both the numerical and the analytical solutions. The error introduced by this method is $E_N \approx 1.36 \times \Delta t$. This indicates a relation that is generally true: The error is smaller for smaller values of $\Delta t$, but the work associated with running through the time steps to compute the numerical solutions is increasing for smaller values of $\Delta t$. We will see plenty of examples of the fact that finer resolutions (i.e., smaller $\Delta t$) means higher accuracy and more work. So life is fair; a low quality solution is cheap and a high quality solution is expensive. Since $E_N$ is proportional to $\Delta t$, we have linear (or first order) convergence. If the error had been proportional to $\Delta t^2$ we would have had quadratic (or second order) convergence, and so on.

## 1.4 Implicit vs. Explicit Numerical Schemes

At this point, we will briefly mention the difference between an implicit and an explicit numerical scheme. When a numerical scheme can be written in the generic form

$$y^{n+1} = F(y^n),$$

like in (1.12), we refer to the scheme as an *explicit scheme* because $y^{n+1}$ can be explicitly computed as a function of $y^n$. Conversely, if an equation has to be solved in order to compute $y^{n+1}$ based on $y^n$, the scheme is referred to as *implicit*. Implicit schemes will be introduced in Chapter 4 of these notes.

## 1.5 But What *Is* a Differential Equation?

We have given you one example of a differential equation, and there are many more in the subsequent chapters. The easiest way to grasp what differential equations are is probably by seeing many examples. However, in general, *differential equations are mathematical relations where the unknown is a function, and derivatives of the function are used in the formulation of the equation.* The solution is a function. Classical mathematical questions related to differential equations are: *Existence:* Is there a solution of the equation? *Uniqueness:* Is there only one solution? *Stability:* Can we slightly perturb the parameters of the equation and obtain almost the same solution? And *Solution:* How can we find the solution? In these notes, our emphasis will be on the last question and we will concentrate on numerical solutions of the equations.

## 1.6 Analytical Solutions

Our focus in these notes will be on numerical methods for solving differential equations. However, we will also encounter some cases where analytical solutions (i.e., solutions given by a formula) can be found. While finding analytical solutions to differential equations is a vast field, these techniques are not typically applicable to the types of problems we will be studying, which is why we will rely on numerical methods. In this section, we will indicate a couple of possible approaches to obtain analytical solutions to illustrate that it is not entirely mysterious, but we will not delve into the methods in these notes, and you can safely skip this subsection and jump to Chapter 2 if you want.

First, we will demonstrate an approach for obtaining an analytical solution for the simple example studied earlier. We start by repeating that the differential equation is given by

$$y'(t) = y(t) \tag{1.16}$$

with the initial condition $y(0) = 1$. To find an analytical solution for this problem, we first write the equation in the form

$$\frac{y'(t)}{y(t)} = 1. \tag{1.17}$$

Now, we can integrate both sides from 0 to $t$ and get

$$\int_0^t \frac{y'(\tau)}{y(\tau)} d\tau = \int_0^t 1 d\tau. \tag{1.18}$$

Therefore,

$$[\ln(y(\tau))]_0^t = t, \tag{1.19}$$

or

$$\ln(y(t)) - \ln(y(0)) = t. \tag{1.20}$$

Since $y(0) = 1$, and thus $\ln(y(0)) = 0$, (1.20) reads

$$\ln(y(t)) = t, \tag{1.21}$$

and raising $e$ to the power of both sides of the equation, we get the analytical solution

$$y(t) = e^t. \tag{1.22}$$

This way of finding the solution of a differential equation can be generalized to equations of the form

$$F'(y)y'(t) = F(y(t)) \tag{1.23}$$

with the initial condition $y(0) = y_0$, where $y_0$ is a given number. Here, $F = F(y)$ is assumed to be some invertible function. By following the steps above we find that

$$[\ln(F(y(\tau)))]_0^t = t, \tag{1.24}$$

or

$$\ln\left(\frac{F(y(t))}{F(y_0)}\right) = t, \tag{1.25}$$

so

$$F(y(t)) = F(y_0)e^t, \tag{1.26}$$

and, finally,

$$y(t) = F^{-1}(F(y_0)e^t). \tag{1.27}$$

## 1.6.1 Integrating Factors

Another way of obtaining analytical solutions of differential equations is by applying *integrating factors*. We can illustrate this technique by considering the equation

$$y'(t) + p(t)y(t) = q(t), \tag{1.28}$$

with the initial condition $y(0) = y_0$, where $y_0$ is given. Here, we assume that $p = p(t)$ and $q = q(t)$ are known functions. In addition, we assume that we have a function $P = P(t)$ with the special property that

$$P'(t) - p(t). \tag{1.29}$$

By multiplying (1.28) by the integrating factor

$$e^{P(t)}$$

we get

$$e^{P(t)}y'(t) + e^{P(t)}p(t)y(t) = e^{P(t)}q(t). \tag{1.30}$$

Here we observe that (because of (1.29)) the left-hand side can be written in a more compact manner as,

$$e^{P(t)}y'(t) + e^{P(t)}p(t)y(t) = (e^{P(t)}y(t))'. \tag{1.31}$$

and therefore (1.30) can be rewritten to read,

$$(e^{P(t)}y(t))' = e^{P(t)}q(t). \tag{1.32}$$

Integrating both sides from 0 to $t$, we obtain,

$$e^{P(t)}y(t) - e^{P(0)}y_0 = \int_0^t e^{P(\tau)}q(\tau)d\tau. \tag{1.33}$$

Finally, the solution is given by

$$y(t) = e^{-P(t)} \left( e^{P(0)} y_0 + \int_0^t e^{P(\tau)} q(\tau) d\tau \right). \tag{1.34}$$

In the case of the simple example equation considered in this chapter, $y'(t) = y(t)$, we note that the equation can be written in the form (1.28) for $p(t) = -1$, $q(t) = 0$, and $y_0 = 1$. Furthermore, (1.29) is fulfilled if $P(t) = -t$. Inserting these functions into (1.34), we obtain the analytical solution

$$y(t) = e^{-(-t)} \left( e^0 \cdot 1 + \int_0^t e^{-\tau} \cdot 0 d\tau \right) = e^t. \tag{1.35}$$

## 1.7 Comments and Further Reading

Here is a list of suggested further reading on a few of the overarching topics of these notes.

1. Throughout these lecture notes, we will refer to results from calculus. A good book on the topic is [12].
2. Introductions to the basics of scientific computing are presented in, e.g., [3, 4, 5, 6, 10, 11, 14, 17].
3. Introductions to differential equations can be found in [7, 8, 13, 16]. There are many texts on analytical solutions of differential equations; one comprehensive collection of solutions is given in [15]. Analytical techniques like the methods described above are well covered in [2].
4. A comprehensive introduction to mathematical biology can be found in [9].
5. For those interested in reading about the place of mathematics in biology, and the role of biology in mathematics, we strongly recommend [1].

## References

[1] Cohen JE (2004) Mathematics is biology's next microscope, only better; biology is mathematics' next physics, only better. PLoS Biology 2(12):e439
[2] Constanda C (2013) Differential Equations: A Primer for Scientists and Engineers. Springer
[3] Deuflhard P, Hohmann A (2003) Numerical analysis in modern scientific computing: an introduction, vol 43. Springer
[4] Heath MT (2018) Scientific computing: an introductory survey, revised second edition. SIAM
[5] Kincaid D, Kincaid DR, Cheney EW (2009) Numerical analysis: mathematics of scientific computing, vol 2. American Mathematical Society
[6] Langtangen HP, Langtangen HP (2011) A primer on scientific programming with Python, vol 1. Springer

[7] Logan JD (2014) Applied partial differential equations. Springer
[8] Martin B (1993) Differential equations and their applications: an introduction to applied mathematics. Springer
[9] Murray JD (2002) Mathematical biology: I. An introduction. Springer
[10] Quarteroni A, Saleri F, Gervasio P (2006) Scientific computing with MATLAB and Octave, vol 3. Springer
[11] Shen W (2019) An introduction to numerical computation. World Scientific
[12] Stewart J (2015) Calculus. Cengage Learning
[13] Strang G (2014) Differential equations and linear algebra. Wellesley-Cambridge Press Wellesley
[14] Sundnes J (2020) Introduction to scientific programming with Python. Springer
[15] Tenenbaum M, Pollard H (1985) Ordinary differential equations: an elementary textbook for students of mathematics, engineering, and the sciences. Courier Corporation
[16] Tveito A, Winther R (2009) Introduction to partial differential equations; a computational approach, 2nd edn. Springer
[17] Tveito A, Langtangen HP, Nielsen BF, Cai X (2010) Elements of scientific computing. Springer

# Chapter 2
# A System of Ordinary Differential Equations

Above, we introduced a very simple differential equation given by

$$y'(t) = y(t). \tag{2.1}$$

This is referred to as an *ordinary differential equation (ODE)* because it involves the derivative with respect to only one variable. If the derivative with respect to more than one variable is involved, the equation is referred to as a *partial differential equation (PDE)*. We will come back to PDEs in Chapter 3, and spend some time discussing how to solve them, but in this chapter, we will keep focusing on ODEs.

The ODE (2.1) is a *scalar* equation because there is only a single unknown function to be found. Now, we will start considering *systems* of ODEs. A typical system of ODEs can be written in the form

$$y'(t) = F(y(t)). \tag{2.2}$$

Here, $y$ is a vector and $F$ is vector valued function.

## 2.1 The FitzHugh-Nagumo Model

We will introduce numerical methods for systems of ODEs by considering the celebrated[1] FitzHugh-Nagumo model published by FitzHugh [1] in 1961 and, independently, by Nagumo et. al. [2] in 1962. The model is a system of ordinary differential equations with two unknowns, and is commonly used as a simple model for the action potentials of excitable pacemaker cells.

We consider the following version of the FitzHugh-Nagumo model,

$$v' = c_1 v(v - a)(1 - v) - c_2 w, \tag{2.3}$$
$$w' = b(v - dw). \tag{2.4}$$

---

[1] These two papers together are cited more than 11,000 times.

© The Author(s) 2023
K. Horgmo Jæger, A. Tveito, *Differential Equations for Studies in Computational Electrophysiology*,
Simula SpringerBriefs on Computing 14, https://doi.org/10.1007/978-3-031-30852-9_2

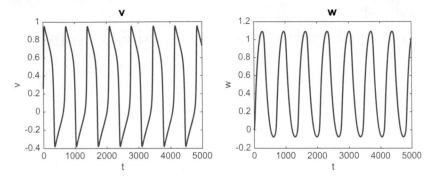

**Fig. 2.1** Numerical solutions $v_n$ (left) and $w_n$ (right) of the FitzHugh-Nagumo model specified by (2.3)–(2.5). The numerical scheme used to compute the solutions is specified in (2.8)–(2.9), and we have used $\Delta t = 1$ and the initial conditions $v_0 = 0.26$ and $w_0 = 0$.

Here, the constants are given by

$$a = -0.12, \; c_1 = 0.175, \; c_2 = 0.03, \; b = 0.011, \; d = 0.55, \tag{2.5}$$

and the unknown functions are $v$ and $w$. In order to solve the system of equations numerically, we use the steps introduced for the scalar equation in Chapter 1 and start by replacing derivatives by differences. The discrete system then reads

$$\frac{v_{n+1} - v_n}{\Delta t} = c_1 v_n (v_n - a)(1 - v_n) - c_2 w_n, \tag{2.6}$$

$$\frac{w_{n+1} - w_n}{\Delta t} = b(v_n - d w_n). \tag{2.7}$$

Again, we reorganize this system to write it in *computational form*,

$$v_{n+1} = v_n + \Delta t [c_1 v_n (v_n - a)(1 - v_n) - c_2 w_n], \tag{2.8}$$

$$w_{n+1} = w_n + \Delta t [b(v_n - d w_n)]. \tag{2.9}$$

This time, however, we note that we will need a piece of software to compute the solutions. But it is straightforward to implement this since the numerical solution at time $t_{n+1}$ is an explicit function of the numerical solution at time $t_n$.

### 2.1.1 Numerical Computations

We assume that the solution is known initially, so we define (for instance) $v_0 = 0.26$ and $w_0 = 0$. Here, it is useful to note that if we put both $v_0$ and $w_0$ equal to zero, the solution will remain zero (for both $v$ and $w$) for all time. But if we perturb $v$ a little, we get very different solutions. In Fig. 2.1, we show the numerical solution from $t = 0$ to $t = T = 5000$. In the computation, we have used $N = 5000$ which

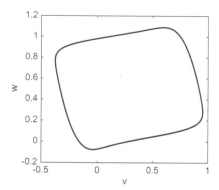

**Fig. 2.2** The numerical
solutions $v_n$ and $w_n$ of the
FitzHugh-Nagumo model
from Fig. 2.1 displayed in a
parametric plot with $v_n$ on the
$x$-axis and $w_n$ on the $y$-axis.

**Table 2.1** Error of the numerical solution of the FitzHugh-Nagumo model specified by (2.3)–(2.5) at $t = T = 5000$ for different values of $\Delta t = \frac{T}{N}$. The error is defined as $E_N = |v - v_N| + |w - w_N|$, where $v$ and $w$ are the numerical solutions for a very fine resolution ($\Delta t = 0.001$), and $v_N$ and $w_N$ are the numerical solutions for larger values of $\Delta t$.

| $N$ | $\Delta t$ | $E_N$ | $E_N / \Delta t$ |
|---|---|---|---|
| 500 | 10 | 0.0923 | 0.0092 |
| 1000 | 5 | 0.0433 | 0.0087 |
| 5000 | 1 | 0.00727 | 0.0073 |
| 10000 | 0.5 | 0.00353 | 0.0071 |
| 50000 | 0.1 | 0.000682 | 0.0068 |

gives $\Delta t = T/N = 1$. In Fig. 2.2, we show the numerical solutions $(v_n, w_n)_{n=1}^{N}$ in a parametric plot, and we note that the solutions are periodic. In electrophysiology, such solutions are useful for studying pacemaker cells that keep on creating action potentials at a steady rate.

## 2.2 What Is the Error?

In the very simple equation in the previous chapter, we had a formula for the analytical solution and a formula for the numerical solution, and thus it was straightforward to find the error introduced by replacing a derivative by a difference. For the FitzHugh-Nagumo equations, this is harder. In numerical analysis there are techniques for proving error bounds for numerical methods. But the proofs tend to be very technical and often involve constants that need to be estimated. So we are looking for something simpler. If we just assume that we have convergence towards the correct solution as $\Delta t$ tends to zero, we can compute an accurate approximation of the solution by using a very small $\Delta t$ and then monitoring the convergence towards this highly resolved solution.

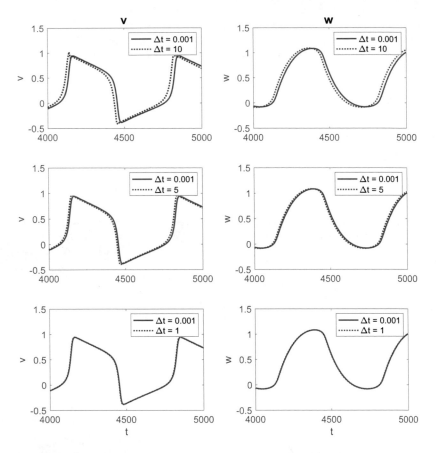

**Fig. 2.3** Numerical solutions $v_n$ (left) and $w_n$ (right) of the FitzHugh-Nagumo model specified by (2.3)–(2.5). We compare the solutions for a very fine resolution ($\Delta t = 0.001$, solid blue line) to the solutions for three cases of coarser resolution (dotted orange line). In the upper panel, we consider $\Delta t = 10$, in the middle panel, we consider $\Delta t = 5$, and in the lower panel, we consider $\Delta t = 1$. To improve visibility, we only show the solutions from $t = 4000$ to $t = 5000$. As $\Delta t$ is decreased, the coarse numerical solutions are more similar to the numerical solution computed with a very small $\Delta t$, and for $\Delta t = 1$ the two solutions are indistinguishable.

For simplicity, we assume that we are merely interested in the error at the final time $t = T = 5000$. We first compute the solution using an extremely fine resolution ($\Delta t = 0.001$) and regard that as the correct solution at time $T$. Then, we compute solutions for varying resolutions (different values of the time step $\Delta t$) and compare the "correct" and approximate solutions. In Fig. 2.3, we compare the solutions for a few different choices of $\Delta t$. The error defined by $E_N = |v - v_N| + |w - w_N|$, where $(v, w)$ denotes the fine scale solution, is given in Table 2.1. Again, we observe that $E_N/\Delta t$ is more or less constant and we therefore conclude again that the error in the numerical solution is proportional to the time step $\Delta t$. In other words, the convergence is linear (or first order) in $\Delta t$.

**Fig. 2.4** Illustration of a definition of the action potential duration, APD50. The APD50 value is defined as the duration between the two times $v$ crosses the threshold value defined at the center between the maximum and minimum value of $v$.

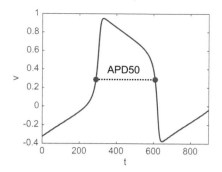

## 2.3 Upstroke Velocity and Action Potential Duration

In applications of the FitzHugh-Nagumo model, the unknown function $v$ is often used to represent the membrane potential of an excitable cell, e.g., a neuron or a cardiac cell, firing a sequence of action potentials. A single action potential from the solution of the FitzHugh-Nagumo model is illustrated in Fig. 2.4. In general terms, the action potential first consists of a period during which the value of $v$ increases slowly, followed by a more rapid increase (the upstroke). Then, $v$ decreases relatively slowly for a while before it decreases rapidly back to the minimum value and starts increasing again. In this setting, we often refer to $v$ increasing as *depolarization*, and $v$ decreasing as *repolarization*. We will come back to these terms below where we introduce models with proper physical units.

By solving the FitzHugh-Nagumo model equations for different values of the model constants, or parameters, $(a, c_1, c_2, b,$ and $d)$, we could gain some insight into how the parameters affect the firing of action potentials. For example, we could investigate how the parameters affect the frequency of firing or the shape of the fired action potentials. Two properties that are of interest regarding the shape of the action potential are the *maximal upstroke velocity* and the *action potential duration*. The maximal upstroke velocity is often defined as the maximum value of the derivative of $v$ with respect to time. Using a finite difference approximation of the derivative, this can be defined as

$$\max_n \left( \frac{v_{n+1} - v_n}{\Delta t} \right). \tag{2.10}$$

The action potential duration is often defined in terms of a given percentage of repolarization, for example APD50 or APD90, for 50% or 90% repolarization, respectively. Here, APD50 represents the duration from the start of the action potential until the membrane potential reaches a value that is 50% repolarized (i.e., at $v_{50} = 0.5\,(\max_n(v_n) + \min_n(v_n))$), at $t_{50}^{\text{down}}$. Similarly, APD90 is defined as the duration from the start of the action potential until the membrane potential reaches a value that is 90% repolarized (i.e., at $v_{90} = \max_n(v_n) - 0.9(\max_n(v_n) - \min_n(v_n))$), at $t_{90}^{\text{down}}$. The start of the action potential used in the definition of the action potential duration can, for example, be defined as the point in time when the maximal

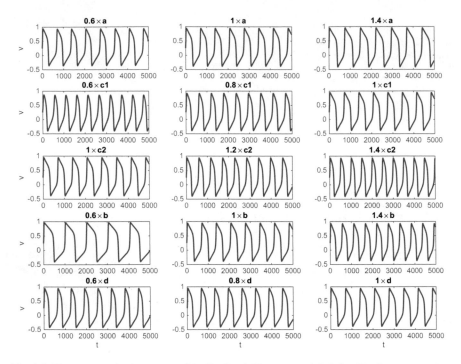

**Fig. 2.5** The numerical solution, $v_n$, of the FitzHugh-Nagumo model, defined by the two equations $v' = c_1 v(v - a)(1 - v) - c_2 w$, and $w' = b(v - dw)$. The parameters are as specified in (2.5), except that in each row of the figure, the value of either $a, c_1, c_2, b,$ or $d$ is adjusted. The title above each plot specifies the parameter change.

upstroke velocity occurs, or when the membrane potential crosses the 50% or 90% repolarization thresholds, $v_{50}$ or $v_{90}$, during the upstroke, denoted by $t_{50}^{\text{up}}$ or $t_{90}^{\text{up}}$, respectively. In the latter case, APD50 and APD90 can be defined as

$$\text{APD50} = t_{50}^{\text{down}} - t_{50}^{\text{up}}, \tag{2.11}$$

$$\text{APD90} = t_{90}^{\text{down}} - t_{90}^{\text{up}}. \tag{2.12}$$

Such a definition of APD50 is illustrated in Fig. 2.4.

In Fig. 2.5, we show the numerical solution, $v_n$, of the FitzHugh-Nagumo model with different choices of parameters. In each row, we consider three different values of one of the parameters $a, c_1, c_2, b,$ or $d$, and keep the remaining values fixed at the values specified in (2.5). In the plots, we observe that increasing the value of $c_1$ appears to make the action potentials longer and the firing frequency slower, whereas the opposite effect is observed when $c_2$ or $b$ are increased. In Fig. 2.6, we study the effects on the individual action potentials more closely. In the left panel, we have zoomed in on the points in time representing the action potential upstroke. We observe that decreasing the value of $c_1$ reduces the upstroke velocity, but changing the other parameters do not seem to have a significant effect on the upstroke. In the

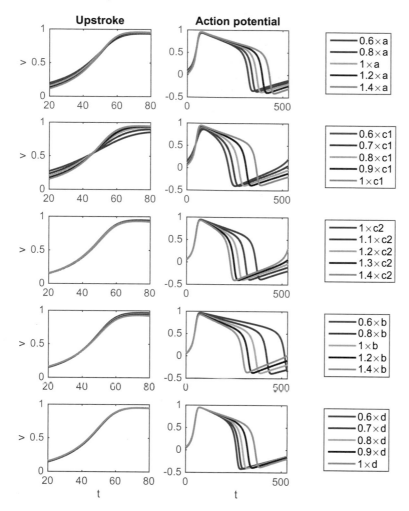

**Fig. 2.6** The numerical solution, $v_n$, of the FitzHugh-Nagumo model, defined by the two equations $v' = c_1 v(v-a)(1-v) - c_2 w$, and $w' = b(v-dw)$. The parameters are as specified in (2.5), except that in each row, the value of either $a, c_1, c_2, b$, or $d$ is adjusted. The legends at the right-hand side of each row specify the parameter changes. The time axes of the solutions are adjusted such that the maximal upstroke velocity occurs at the same time for all the parameter changes. The left panel shows the upstroke of the action potential and the right panel shows one action potential for each parameter set.

right panel, we consider a single action potential for the different parameter choices. We observe that all the parameters have a significant effect on the action potential duration.

# References

[1] FitzHugh R (1961) Impulses and physiological states in theoretical models of nerve membrane. Biophysical Journal 1(6):445–466

[2] Nagumo J, Arimoto S, Yoshizawa S (1962) An active pulse transmission line simulating nerve axon. Proceedings of the IRE 50(10):2061–2070

# Chapter 3
# The Diffusion Equation

The diffusion equation appears in many applications in science and engineering, and computational physiology is no exception. In its most basic form, the diffusion equation is also useful as an example of how to deal with a PDE using numerical methods. We will start by considering it as a stand-alone model, but in the next chapters we will study it in combination with non-linear ODEs. This chapter therefore serves as a warm-up for the more complex models. We will also follow the path we started above. In the very simplest case of an ODE, we found a formula for the solution of both the differential equation and the numerical scheme approximating the equation. One nice consequence of this is that, as we saw above, we can explicitly study the error introduced by the numerical approximation. In this chapter, we will use the same approach to study the error of the numerical approximation of the diffusion equation.

## 3.1 The Problem = Equation + Initial Values + Boundary Conditions

We consider the diffusion equation

$$\frac{\partial u}{\partial t}(t, x) = \frac{\partial^2 u}{\partial x^2}(t, x). \tag{3.1}$$

Here, $u$ models[1] the concentration of a chemical in a spatial domain, $x$ is the spatial variable and $t$ is time. And since partial derivatives are present, the equation is a PDE (see page 13). We study the problem for values of $x$ in the spatial domain $(0, 1)$ and for time $t$ in the interval $(0, T]$. We will assume that the solution is given by $u^0(x)$ at time $t = 0$, and that $u = 0$ at $x = 0$ and $x = 1$ for all time in the interval $[0, T]$.

---

[1] In the beginning of these notes, we treat all models as unitless. In models of electrophysiology, units can be a true nightmare so we follow the prudent path of trying to deal with one nightmare at the time. Units will come later.

© The Author(s) 2023
K. Horgmo Jæger, A. Tveito, *Differential Equations for Studies in Computational Electrophysiology*,
Simula SpringerBriefs on Computing 14, https://doi.org/10.1007/978-3-031-30852-9_3

The function $u^0 = u^0(x)$ is called the *initial condition*, and $u = 0$ at $x = 0$ and $x = 1$ are called *boundary conditions*. In particular, these boundary conditions are referred to as Dirichlet boundary conditions, as opposed to Neumann boundary conditions, which will be introduced below.

We will consider the special case of

$$u^0(x) = \sin(\pi x). \tag{3.2}$$

For that particular initial condition, the analytical solution of the diffusion equation (3.1) is given by

$$u(t, x) = e^{-\pi^2 t} \sin(\pi x). \tag{3.3}$$

This can be verified by observing that the boundary conditions are satisfied since clearly $u(t, 0) = u(t, 1) = 0$, since $\sin(0) = 0$, and the initial condition is also satisfied since $u(0, x) = \sin(\pi x) = u^0(x)$. It remains to be seen whether equation (3.1) is satisfied by $u(t, x)$. To this end, we note that

$$\frac{\partial u}{\partial t} = -\pi^2 e^{-\pi^2 t} \sin(\pi x) = -\pi^2 u, \tag{3.4}$$

and

$$\frac{\partial^2 u(t, x)}{\partial x^2} = -\pi^2 e^{-\pi^2 t} \sin(\pi x) = -\pi^2 u, \tag{3.5}$$

so (3.1) holds.

## 3.2  The Numerical Scheme

We define a numerical approximation of the problem simply by replacing derivatives by differences. In order to replace the left-hand side of (3.1), we use the same formula as above and state that

$$\frac{\partial u}{\partial t} \approx \frac{u(t + \Delta t, x) - u(t, x)}{\Delta t}. \tag{3.6}$$

To approximate the right-hand side of (3.1), we need to recall from calculus that the Taylor series of $u$ states that

$$u(t, x + \Delta x) \approx u(t, x) + \Delta x u_x(t, x) + \frac{1}{2} \Delta x^2 u_{xx}(t, x), \tag{3.7}$$

where $u_x$ and $u_{xx}$ are shorthand for $\frac{\partial u(t,x)}{\partial x}$ and $\frac{\partial^2 u(t,x)}{\partial x^2}$, respectively. Similarly,

$$u(t, x - \Delta x) \approx u(t, x) - \Delta x u_x(t, x) + \frac{1}{2} \Delta x^2 u_{xx}(t, x), \tag{3.8}$$

and therefore, by adding (3.7) and (3.8), we get the approximation,

$$u_{xx}(t, x) \approx \frac{u(t, x - \Delta x) - 2u(t, x) + u(t, x + \Delta x)}{\Delta x^2}. \tag{3.9}$$

As above, we introduce $t_n = n \times \Delta t$, where $\Delta t = T/N$ for a sufficiently large integer $N$. Furthermore, we introduce the spatial mesh points given by $x_j = (j - 1) \times \Delta x$ for $j = 1, ..., M$, where $\Delta x = 1/(M - 1)$ for a suitable[2] integer $M$.

We have defined the points $\{(t_n, x_j)\}$ and we let $u_j^n$ denote a numerical approximation in these points. It remains to define these values. The beginning is very simple; we clearly want the numerical scheme to satisfy the initial condition so we define,

$$u_j^0 = u^0(x_j), \tag{3.10}$$

and for our special problem we have $u_j^0 = \sin(\pi x_j)$. Next, we apply the boundary conditions and define

$$u_1^n = 0, \tag{3.11}$$

and

$$u_M^n = 0, \tag{3.12}$$

for all $n \leq N$. We still need to find values of $u_j^n$ for $j = 2, \ldots, M-1$ and $1 < n \leq N$. We find these by using the finite difference approximations (3.6) and (3.9). Specifically, we define $u_j^n$ by the following numerical scheme,

$$\frac{u_j^{n+1} - u_j^n}{\Delta t} = \frac{u_{j-1}^n - 2u_j^n + u_{j+1}^n}{\Delta x^2}. \tag{3.13}$$

By defining

$$\rho = \frac{\Delta t}{\Delta x^2}, \tag{3.14}$$

we can write the scheme in computational form,

$$u_j^{n+1} = \rho u_{j-1}^n + (1 - 2\rho)u_j^n + \rho u_{j+1}^n. \tag{3.15}$$

Starting with $n = 0$, we note that we can compute all the values at time $t_1 = \Delta t$ since $u_j^1$ only depends on values at the time $t_0 = 0$ and they are given by the initial condition $u_j^0$ ($j = 1, \ldots, M$). When the values at time $t_1 = \Delta t$ have been computed, we can use them to compute all the values at $t = 2\Delta t$ and so forth. The scheme is illustrated in Fig. 3.1.

## 3.3 Analytical Solution of the Numerical Scheme and Error

This subsection is very technical and can be skipped without serious consequences (jump to Section 3.4). The purpose of the section is to provide formulas for the

---

[2] 'Sufficiently large integer' and 'suitable integer' are intentionally vague – stay calm – it will become clear (clearer) later on.

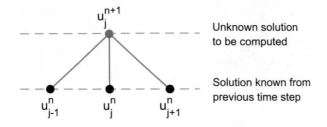

**Fig. 3.1** Illustration of the explicit finite difference scheme for the diffusion equation. The solutions $u_{j-1}^n$, $u_j^n$ and $u_{j+1}^n$ from the previous time step are used to compute the new solution, $u_j^{n+1}$, at this time step. The scheme is given by (3.15).

numerical solution and the associated error for a PDE like we did for an ODE in (1.13). Such formulas are available only in rare cases, but can be useful for analyzing the error of the numerical approximation.

For a general initial condition, we have to write a program to use the scheme (3.15), and we will do that below, but for the special initial condition (3.2), we can find a formula for the numerical solution. It is given by

$$u_j^n = (1 - \Delta t \mu)^n \sin(\pi x_j), \tag{3.16}$$

where

$$\mu = \frac{4}{\Delta x^2} \sin^2(\pi \Delta x / 2). \tag{3.17}$$

In order to show that this is the solution, we first check that the formula holds at time $t_0$. Inserting $n = 0$ into (3.16), we get $u_j^0 = \sin(\pi x_j)$, which matches the numerical solution at time $t_0 = 0$ (see (3.2) and (3.10)). Next, we assume that the formula is correct at time $t_n$ and show that it also holds at time $t_{n+1}$, – then the formula holds by induction. By inserting (3.16) into (3.15), we get,

$$u_j^{n+1} = \rho u_{j-1}^n + (1 - 2\rho) u_j^n + \rho u_{j+1}^n \tag{3.18}$$
$$= (1 - \Delta t \mu)^n [\rho \sin(\pi x_{j-1}) + (1 - 2\rho) \sin(\pi x_j) + \rho \sin(\pi x_{j+1})]. \tag{3.19}$$

We need two trigonometric identities from calculus to proceed from here,

$$1 - \cos(y) = 2 \sin^2(y/2), \tag{3.20}$$

and

$$\sin(x + y) + \sin(x - y) = 2 \cos(y) \sin(x). \tag{3.21}$$

These identities hold for all values of $x$ and $y$. We define

$$z_j = \rho \sin(\pi x_{j-1}) + (1 - 2\rho) \sin(\pi x_j) + \rho \sin(\pi x_{j+1}) \tag{3.22}$$

and rewrite it like

$$z_j = \sin(\pi x_j) + \rho[\sin(\pi x_{j-1}) + \sin(\pi x_{j+1}) - 2\sin(\pi x_j)] \tag{3.23}$$
$$= \sin(\pi x_j) + \rho[\sin(\pi x_j - \pi\Delta x) + \sin(\pi x_j + \pi\Delta x) - 2\sin(\pi x_j)], \tag{3.24}$$

so by (3.21), we get

$$z_j = \sin(\pi x_j) + \rho[2\cos(\pi\Delta x)\sin(\pi x_j) - 2\sin(\pi x_j)] \tag{3.25}$$
$$= \sin(\pi x_j) + 2\rho[\cos(\pi\Delta x) - 1]\sin(\pi x_j). \tag{3.26}$$

By (3.20), we now get,

$$z_j = \sin(\pi x_j) - 4\rho\sin^2(\pi\Delta x/2)\sin(\pi x_j) \tag{3.27}$$
$$= \sin(\pi x_j)[1 - 4\rho\sin^2(\pi\Delta x/2)], \tag{3.28}$$

so, by (3.14) and (3.16), we have

$$z_j = [1 - \Delta t\mu]\sin(\pi x_j). \tag{3.29}$$

Here, $\mu$ is defined by (3.17). By combining (3.19), (3.22) and (3.29), we find that

$$u_j^{n+1} = (1 - \Delta t\mu)^{n+1}\sin(\pi x_j), \tag{3.30}$$

and thus the formula (3.16) holds by induction.

### 3.3.1 What Is the Error?

Since we now have the analytical solution of the PDE (3.1) given by

$$u(t, x) = e^{-\pi^2 t}\sin(\pi x) \tag{3.31}$$

and a formula (3.16) for the numerical solution

$$u_j^n = (1 - \Delta t\mu)^n\sin(\pi x_j) \tag{3.32}$$

it is straightforward to compare the analytical and numerical solutions. The spatial part of the solutions are identical, and therefore it is sufficient to consider the temporal part of the solution. We thus consider the relative error in time defined by

$$E_n = \frac{|(1 - \Delta t\mu)^n - e^{-\pi^2 t_n}|}{e^{-\pi^2 t_n}}. \tag{3.33}$$

In Table 3.1 we show $E_N$ where $N = 1/\Delta t$ for several values of $\Delta t$. We also show $E_N/\Delta t$ and we see that, again, this is more or less constant and therefore we again have linear convergence with $E_N \approx 48 \times \Delta t$.

**Table 3.1** Error of the temporal part of the numerical solution of the diffusion equation at $t = T = 1$ for $\Delta x = 0.001$ and different values of $\Delta t = \frac{T}{N}$. The error is defined in (3.33).

| $N$ | $\Delta t$ | $E_N$ | $E_N/\Delta t$ |
|---|---|---|---|
| 100 | 0.01 | 0.406 | 40.6 |
| 1000 | 0.001 | 0.0478 | 47.8 |
| 10000 | 0.0001 | 0.00485 | 48.5 |
| 100000 | 0.00001 | 0.000479 | 47.9 |

### 3.3.2 More on the Error

We can go one step further in comparing the analytical (3.31) and numerical (3.32) solutions using the formulas for these solutions. Since the spatial dependency of the solutions are equal, we just pick $x = 1/2$ and then we want to compare the analytical solution

$$U(T) = u(T, 1/2) = e^{-\pi^2 T}. \tag{3.34}$$

and the numerical solution

$$U^N = u^N_{(M+1)/2} = (1 - \Delta t \mu)^N, \tag{3.35}$$

at the final time $t = T = N \times \Delta t$. Here we have assumed that $M$ is an odd number. Recall, again from calculus, that the Taylor series of the sine function states that[3]

$$\sin(z) = z + O(z^3), \tag{3.36}$$

so for small values of $z$, we have $\sin(z) \approx z$. We can use this to approximate the value of

$$\mu = \frac{4}{\Delta x^2} \sin^2(\pi \Delta x/2) \tag{3.37}$$

for small values of $\Delta t$. By the Taylor series, we get

$$\mu \approx \frac{4}{\Delta x^2}(\pi \Delta x/2)^2 = \pi^2. \tag{3.38}$$

So the numerical solution satisfies

$$U^N \approx (1 - \Delta t \pi^2)^N. \tag{3.39}$$

We can also approximate the analytical solution using the following Taylor series for the exponential function,

---

[3] You don't remember the $O$-notation? It is shorthand for telling how fast something goes to zero. So $f(x) = O(x^2)$ means that $f(x)$ goes to zero as fast as $x^2$ goes to zero for small values of $x$. In numerical analysis this notation is often used to indicate the size of an error term. Often, it is the remainder of a truncated Taylor series.

$$e^{-z} = 1 - z + O(z^2). \tag{3.40}$$

By using this approximation, we find that the analytical solution at $x = 1/2$ can be approximated by

$$U(T) = e^{-\pi^2 T} = e^{-\pi^2 N \Delta t} = (e^{-\pi^2 \Delta t})^N \approx (1 - \pi^2 \Delta t)^N, \tag{3.41}$$

hence, clearly, we have

$$U^N \approx U(T). \tag{3.42}$$

## 3.4 Instabilities in the Numerical Solution

Numerical instabilities often appear and it can be hard to understand the origin. Here, we will show instabilities that appear because of too long time steps, but keep in mind that a long list of other instabilities can appear as well. Differential equations can have oscillatory solutions, but as the mesh parameters are refined, the numerical solution should converge towards the correct solution. Numerical instabilities, on the other hand, manifest themselves by diverging as the mesh is refined, rather than converging.

### 3.4.1 Specification of a Stable Problem with Unstable Numerical Solutions

Suppose you have a tank of length 1 filled with water, where you have ink added to the water for $x \leq 1/2$ but no ink for $x > 1/2$. In the middle of the tank, at $x = 1/2$, there is an impermeable membrane, so there is no ink leaking from the left to the right side of the tank. At time $t = 0$, we remove the membrane and we are curious about what is going to happen. Intuitively, we expect the ink to diffuse to the right-hand side of the tank and that, eventually, the ink will be uniformly distributed throughout the tank. Qualitatively, we can get an impression of what happens by solving the diffusion equation

$$\frac{\partial u}{\partial t}(t, x) = \frac{\partial^2 u}{\partial x^2}(t, x). \tag{3.43}$$

We will use the initial condition $u(0, x) = 1$ for $x \leq 1/2$ and $u(0, x) = 0$ for $x > 1/2$. The boundary conditions are given

$$\frac{\partial u}{\partial x}(t, 0) = \frac{\partial u}{\partial x}(t, 1) = 0, \tag{3.44}$$

i.e., that the spatial derivative is zero at the boundary. This is referred to as a no-flux boundary condition, or a Neumann boundary condition, and simply means that we don't allow the ink to leak out of the tank.

### 3.4.2 Numerical Scheme for Neumann Boundary Conditions

In order to deal with this alternative set of boundary conditions in the numerical scheme, we consider the two Taylor series of $u$ considered in Section 3.2, i.e.,

$$u(t, x + \Delta x) \approx u(t, x) + \Delta x u_x(t, x) + \frac{1}{2} \Delta x^2 u_{xx}(t, x), \tag{3.45}$$

and

$$u(t, x - \Delta x) \approx u(t, x) - \Delta x u_x(t, x) + \frac{1}{2} \Delta x^2 u_{xx}(t, x). \tag{3.46}$$

By subtracting (3.46) from (3.45), we obtain

$$u(t, x + \Delta x) - u(t, x - \Delta x) \approx 2\Delta x u_x(t, x), \tag{3.47}$$

which yields

$$u_x(t, x) \approx \frac{u(t, x + \Delta x) - u(t, x - \Delta x)}{2\Delta x}. \tag{3.48}$$

Replacing the derivatives by this difference in the boundary condition (4.24), we get

$$\frac{u_2^n - u_0^n}{2\Delta x} = 0, \qquad \frac{u_{M+1}^n - u_{M-1}^n}{2\Delta x} = 0, \tag{3.49}$$

which yield

$$u_0^n = u_2^n, \qquad u_{M+1}^n = u_{M-1}^n, \tag{3.50}$$

for all $n \leq N$. Inserting (3.50) into the main scheme (3.15) for $j = 1$ and $j = M$, we obtain

$$u_1^{n+1} = (1 - 2\rho)u_1^n + 2\rho u_2^n, \tag{3.51}$$

$$u_M^{n+1} = (1 - 2\rho)u_M^n + 2\rho u_{M-1}^n. \tag{3.52}$$

### 3.4.3 Example of Instabilities in the Numerical Solution

In Fig. 3.2 we have solved the problem numerically using $\Delta t = 0.0001$ and $\Delta x = 0.02$ and we note that the solution seems to evolve as we expected. But in Fig. 3.3, we attempt to take longer time steps, and then we see pretty wild oscillations in the numerical solution. This is a classical type of numerical instability that often appears in numerical methods. Usually it helps to reduce the time steps, and it certainly helps in this case.

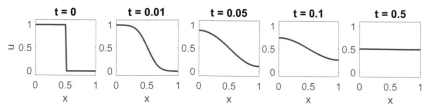

**Fig. 3.2** Numerical solution of the diffusion equation with boundary conditions $\frac{\partial u}{\partial x}(t, 0) = \frac{\partial u}{\partial x}(t, 1) = 0$ and initial conditions $u(0, x) = 1$ for $x \leq 1/2$ and $u(0, x) = 0$ for $x > 1/2$ using $\Delta t = 0.0001$ and $\Delta x = 0.02$. The solution is plotted at five different time points.

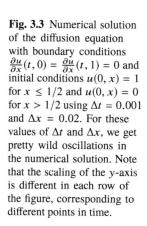

**Fig. 3.3** Numerical solution of the diffusion equation with boundary conditions $\frac{\partial u}{\partial x}(t, 0) = \frac{\partial u}{\partial x}(t, 1) = 0$ and initial conditions $u(0, x) = 1$ for $x \leq 1/2$ and $u(0, x) = 0$ for $x > 1/2$ using $\Delta t = 0.001$ and $\Delta x = 0.02$. For these values of $\Delta t$ and $\Delta x$, we get pretty wild oscillations in the numerical solution. Note that the scaling of the $y$-axis is different in each row of the figure, corresponding to different points in time.

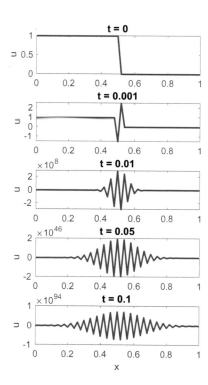

## 3.5 Stability Condition for the Numerical Scheme

The diffusion equation satisfies a maximum principle stating that the solution will
always be bounded by the values of the initial condition and the boundary conditions.
We will show that the numerical solution generated by the scheme (3.15) satisfies
the same principle, provided that the time step is sufficiently small. For simplicity
we again assume that the boundary conditions are given by $u = 0$ for $x = 0$ and
$x = 1$. Concretely, we consider the scheme

$$u_j^{n+1} = \rho u_{j-1}^n + (1 - 2\rho)u_j^n + \rho u_{j+1}^n, \tag{3.53}$$

where we recall that

$$\rho = \frac{\Delta t}{\Delta x^2}, \tag{3.54}$$

and where the boundary conditions are given by $u_1^n = 0$ and $u_M^n = 0$. Define

$$u_+ = \max_j |u_j^0|, \tag{3.55}$$

i.e., the biggest (in absolute value) initial value, and assume that $\rho \leq 1/2$, i.e. we
assume that

$$\Delta t \leq \frac{\Delta x^2}{2}. \tag{3.56}$$

Now, we want to prove that

$$|u_j^n| \leq u_+ \tag{3.57}$$

for all $j \in [2, M - 1]$ and $n \geq 0$. We will show this by induction and start by assuming
that (3.57) holds for an arbitrary value of $n$. Then, by the scheme (3.53), we get[4]

$$
\begin{aligned}
|u_j^{n+1}| &= |\rho u_{j-1}^n + (1 - 2\rho)u_j^n + \rho u_{j+1}^n| \\
&\leq \rho|u_{j-1}^n| + (1 - 2\rho)|u_j^n| + \rho|u_{j+1}^n| \\
&\leq \rho u_+ + (1 - 2\rho)u_+ + \rho u_+ \\
&= u_+
\end{aligned}
$$

and therefore (3.57) holds by induction.

We also note that the criterion (3.56) is in agreement with the results observed in
the numerical example in the previous section. In that case, we had $\frac{\Delta x^2}{2} = 0.0002$.
Thus, for the stable case of $\Delta t = 0.0001$ (Fig. 3.2), the stability criterion (3.56)
was satisfied, whereas for the unstable case of $\Delta t = 0.001$ (Fig. 3.3), the stability
criterion was not satisfied.

When using explicit finite difference schemes to solve equations where the
diffusion equation is part of the problem, a stability condition similar to (3.56)
usually has to be satisfied in order to obtain proper numerical results.

---

[4] Here we use the *triangle equality* stating that $|a + b| \leq |a| + |b|$ for any numbers $a$ and $b$.

## 3.6 A Brief Comment on Uniqueness

We have seen that a formula can be obtained for the solution of the diffusion equation and for the numerical approximation of the same problem. But are these solutions unique? Can there be other solutions than those given by the formulas? In order to see that the solution of the continuous problem is in fact unique, we assume that we have two solutions, $u$ and $v$, with coinciding initial condition given by $f = f(x)$, and the usual Dirichlet boundary conditions. Since $u$ and $v$ solve

$$\frac{\partial u}{\partial t}(t, x) = \frac{\partial^2 u}{\partial x^2}(t, x) \tag{3.58}$$

and

$$\frac{\partial v}{\partial t}(t, x) = \frac{\partial^2 v}{\partial x^2}(t, x), \tag{3.59}$$

respectively, we find (by subtracting (3.59) from (3.58)) that the difference between the solutions, given by $e = u - v$, solves the equation

$$\frac{\partial e}{\partial t}(t, x) = \frac{\partial^2 e}{\partial x^2}(t, x). \tag{3.60}$$

Now, we can define a measure of the difference between these solutions as a function of time,

$$E(t) = \frac{1}{2} \int_0^1 e^2(t, x) dx, \tag{3.61}$$

and observe that

$$E'(t) = \int_0^1 e(t, x) \frac{\partial e}{\partial t}(t, x) dx = \int_0^1 e(t, x) \frac{\partial^2 e}{\partial x^2}(t, x). \tag{3.62}$$

By using integrations by parts, we find that

$$\int_0^1 e(t, x) \frac{\partial^2 e}{\partial x^2}(t, x) = - \int_0^1 \left( \frac{\partial e}{\partial x}(t, x) \right)^2 dx, \tag{3.63}$$

and therefore,

$$E'(t) \le 0. \tag{3.64}$$

Since $u$ and $v$ have the same initial condition given by $f = f(x)$, we clearly have $e(0, x) = 0$ for all relevant values of $x$. Therefore, since $E(0) = 0$, and $E'(t) \le 0$ we have $E(t) \equiv 0$ for all time and thus $u$ and $v$ are equal, and the solution of the problem must be unique. A similar argument can be given for the discrete case, so the numerical solution given by the formula (3.16) is also unique. This is a classical energy argument and it can be extended to also provide stability estimates of the solutions; see, e.g., [1].

# References

[1] Tveito A, Winther R (2009) Introduction to partial differential equations; a computational approach, 2nd edn. Springer

# Chapter 4
# Implicit Numerical Methods

In the previous chapter, we saw that the simple explicit numerical scheme resulted in an instability problem. We also saw that the problem could be resolved by using sufficiently short time steps. But in many situations, short time steps become exceedingly short, as can be seen, e.g., in the stability criterion (3.56). This means that we have to perform computations for a very large number of time steps to reach the final time and it is therefore tempting to look for alternatives. The most common alternative is to use an implicit scheme, which generally allows for much longer time steps.

## 4.1 Explicit and Implicit Numerical Schemes

In Section 1.4 (page 7) we briefly introduced the concept of *explicit* and *implicit* numerical schemes. So far we have only considered explicit schemes and the reason for that is just simplicity. If we have a differential equation that can be written in the form

$$u'(t) = F(u(t)), \qquad (4.1)$$

we can, as seen in Chapter 1, use the the approximation

$$u'(t) \approx \frac{u(t + \Delta t) - u(t)}{\Delta t}, \qquad (4.2)$$

to replace (4.1) by the difference equation,

$$\frac{u^{n+1} - u^n}{\Delta t} = F(u^n), \qquad (4.3)$$

leading to the *explicit scheme*

$$u^{n+1} = u^n + \Delta t F(u^n). \qquad (4.4)$$

K. Horgmo Jæger, A. Tveito, *Differential Equations for Studies in Computational Electrophysiology*, Simula SpringerBriefs on Computing 14, https://doi.org/10.1007/978-3-031-30852-9_4

But an alternative, equally plausible, approach is to replace (4.1) by the difference equation

$$\frac{u^{n+1} - u^n}{\Delta t} = F(u^{n+1}),\tag{4.5}$$

which leads to the *implicit scheme*

$$u^{n+1} - \Delta t F(u^{n+1}) = u^n.\tag{4.6}$$

As we shall see below, the advantage of the implicit scheme is that we obtain numerically stable results for longer time steps, but the disadvantage is that we have to solve a potentially nonlinear equation of the form

$$u - \Delta t F(u) = \bar{u}\tag{4.7}$$

at every time step. Here, $\bar{u}$ is known from the previous time step, and $u$ is the unknown that we need to compute.

We will now show how to derive implicit schemes and demonstrate that they are more stable than the explicit versions. But keep in mind that we also have to deal with accuracy. For this purpose we still want the time steps to be reasonably short without being *too* short.

## 4.2 An Implicit Scheme for the Diffusion Equation

Let us first recall how we derived the explicit scheme for the diffusion equation[1],

$$u_t = u_{xx},\tag{4.8}$$

equipped with the initial and boundary conditions,

$$u(0, x) = u^0(x),\tag{4.9}$$
$$u(t, 0) = u(t, 1) = 0.\tag{4.10}$$

As in (4.3), we can approximate this equation by replacing the time derivative by

$$u_t \approx \frac{u(t + \Delta t) - u(t)}{\Delta t}.\tag{4.11}$$

For the right-hand side of (4.8) there are two alternatives (actually many alternatives). We can either approximate the right-hand side by a difference approximation of $u_{xx}(t, x)$ or of $u_{xx}(t + \Delta t, x)$. The first alternative results in the explicit scheme we used above,

---

[1] If you just browse these notes and didn't recognize the notation used here, we repeat that $u_t = \partial u / \partial t$, and $u_{xx} = \partial^2 u / \partial x^2$ – keep on browsing.

$$\frac{u_j^{n+1} - u_j^n}{\Delta t} = \frac{u_{j-1}^n - 2u_j^n + u_{j+1}^n}{\Delta x^2}, \tag{4.12}$$

and the second alternative results in the implicit scheme,

$$\frac{u_j^{n+1} - u_j^n}{\Delta t} = \frac{u_{j-1}^{n+1} - 2u_j^{n+1} + u_{j+1}^{n+1}}{\Delta x^2}. \tag{4.13}$$

In Fig. 4.1 we have illustrated how this scheme works. Here, we realize that there is a fundamental difference between the explicit (see Fig. 3.1) and implicit schemes. The explicit schemes are very simple since every value is simply an explicit function of the values of the previous time step. This is straightforward to implement in software. But the implicit scheme is much more convoluted. In fact, all values at time $t_{n+1}$ are coupled with all other other values at that the same time step. Well, actually, every point is connected to two neighbors, but these again are coupled to their neighbors and so on. So we end up with a linear system of equations.

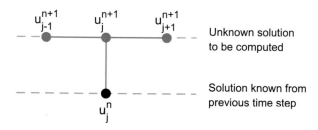

**Fig. 4.1** Illustration of the implicit finite difference scheme for the diffusion equation. The unknown solutions $u_{j-1}^{n+1}$ and $u_{j+1}^{n+1}$ from the present time step, in addition to the known solution $u_j^n$ from the previous time step, are all needed to compute the solution $u_j^{n+1}$. The scheme is given by (4.14), or on matrix form by (4.21).

## 4.3 The Linear System

We will write the scheme (4.13) as a linear system of equations, but first we rearrange it by putting unknowns (variables at time $t_{n+1}$) at the left-hand side and previously computed variables ($t_n$) on the right-hand side of the equation,

$$- \rho u_{j-1}^{n+1} + (1 + 2\rho)u_j^{n+1} - \rho u_{j+1}^{n+1} = u_j^n, \tag{4.14}$$

where again

$$\rho = \frac{\Delta t}{\Delta x^2}. \tag{4.15}$$

The implicit scheme has the form (4.14) for $j = 3, \ldots, M - 2$ but takes a different form for $j = 2$ and $j = M - 1$. Recall that the boundary conditions are specified by (4.10), given by

$$u_1^n = u_M^n = 0 \tag{4.16}$$

for all $n$. Hence, for $j = 2$, the scheme (4.14) takes the form,

$$(1 + 2\rho)u_2^{n+1} - \rho u_3^{n+1} = u_2^n. \tag{4.17}$$

Similarly, for $j = M - 1$, we get the scheme

$$-\rho u_{M-2}^{n+1} + (1 + 2\rho)u_{M-1}^{n+1} = u_{M-1}^n. \tag{4.18}$$

We are now ready to rewrite the somewhat messy scheme defined by (4.14), (4.17), (4.18) in matrix form. To this end, we defined the vectors[2]

$$u^n = \begin{pmatrix} u_2^n \\ u_3^n \\ \vdots \\ u_{M-1}^n \end{pmatrix}, \tag{4.19}$$

and the matrix

$$A = \begin{pmatrix} 1 + 2\rho & -\rho & 0 & \cdots & & 0 \\ -\rho & 1 + 2\rho & -\rho & 0 & \cdots & 0 \\ 0 & -\rho & 1 + 2\rho & -\rho & 0 & \cdots & 0 \\ \vdots & & & \ddots & \ddots & \ddots & \vdots \\ 0 & & \cdots & & 0 & -\rho & 1 + 2\rho \end{pmatrix}. \tag{4.20}$$

With these definitions, we can rewrite the scheme defined by (4.14), (4.17), (4.18) on the simple form

$$Au^{n+1} = u^n. \tag{4.21}$$

The matrix $A$ defined in (4.20) is tridiagonal and this makes the system (4.21) easy to solve. In the software associated these notes, the solution is shown in Matlab, but the system is easy to solve in any numerically oriented computing system.

### 4.3.1 Neumann Boundary Conditions

In Section 3.4 we considered the numerical solution of a specific diffusion equation problem given by

---

[2] Note that $u_1^n$ and $u_M^n$ are given directly by the boundary conditions at every time step, so these values do not have to be included in the vector of unknowns.

$$u_t = u_{xx}, \tag{4.22}$$

with the initial and boundary conditions

$$u(0, x) = \begin{cases} 1, & \text{if } x \le 1/2, \\ 0, & \text{if } x > 1/2, \end{cases} \tag{4.23}$$

$$u_x(t, 0) = u_x(t, 1) = 0. \tag{4.24}$$

For the inner points $j = 2, \ldots, M - 1$, an implicit numerical scheme for this problem can be given by (4.14) like above, but for the boundary points ($j = 1$ and $j = M$), we need to take the different set of boundary conditions into account. One way to do this is by following the same procedure as in Section 3.4.2. That is, using the difference

$$u_x(t, x) \approx \frac{u(t, x + \Delta x) - u(t, x - \Delta x)}{2\Delta x}. \tag{4.25}$$

Replacing the derivatives by this difference in the boundary condition (4.24), we get

$$\frac{u_2^n - u_0^n}{2\Delta x} = 0, \qquad \frac{u_{M+1}^n - u_{M-1}^n}{2\Delta x} = 0, \tag{4.26}$$

which yield

$$u_0^n = u_2^n, \qquad u_{M+1}^n = u_{M-1}^n, \tag{4.27}$$

for all $n \le N$. Inserting (4.27) into the main scheme (4.14) for $j = 1$ and $j = M$, we obtain

$$(1 + 2\rho)u_1^{n+1} - 2\rho u_2^{n+1} = u_1^n, \tag{4.28}$$

$$(1 + 2\rho)u_M^{n+1} - 2\rho u_{M-1}^{n+1} = u_M^n, \tag{4.29}$$

and the scheme can be written on matrix form

$$Au^{n+1} = u^n, \tag{4.30}$$

where

$$u^n = \begin{pmatrix} u_1^n \\ u_2^n \\ \vdots \\ u_M^n \end{pmatrix}, \tag{4.31}$$

and

$$A = \begin{pmatrix} 1 + 2\rho & -2\rho & 0 & \cdots & & 0 \\ -\rho & 1 + 2\rho & -\rho & 0 & \cdots & 0 \\ 0 & -\rho & 1 + 2\rho & -\rho & 0 & \cdots & 0 \\ \vdots & & \ddots & \ddots & \ddots & & \vdots \\ 0 & & \cdots & & 0 & -2\rho & 1 + 2\rho \end{pmatrix}. \tag{4.32}$$

## 4.4 The Implicit Scheme Is Stable

We noticed above that if the time steps we used for the explicit scheme were too long, we got a solution with wild oscillations. In Fig. 4.2 we repeat these computations using the implicit scheme (4.30)–(4.32) and we notice that the oscillations are gone. The computations could indicate that the solutions are stable for any value of $\Delta t$. In fact, unconditional stability of the implicit scheme for the diffusion equation is classical and can be proved. One argument can be found in [3].

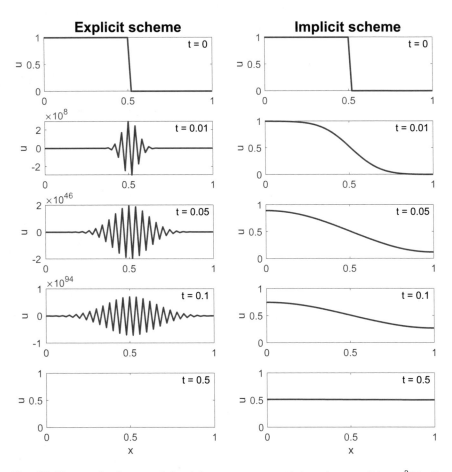

**Fig. 4.2** Numerical solutions of the diffusion equation with boundary conditions $\frac{\partial u}{\partial x}(t, 0) = \frac{\partial u}{\partial x}(t, 1) = 0$ and initial conditions $u(0, x) = 1$ for $x \leq 1/2$ and $u(0, x) = 0$ for $x > 1/2$ using $\Delta t = 0.001$ and $\Delta x = 0.02$. The left panel shows the solution for the explicit scheme (as also seen in Fig. 3.3), and we observe wild oscillations. In the right panel, we show the solution of the implicit scheme, and the solution appears to be more reasonable. Note that at $T = 0.5$, the explicit scheme is so broken down that the solution returned by the scheme are not numbers (NaNs).

## 4.5 Comments and Further Reading

1. As mentioned above, when the PDE under consideration is in one spatial dimension, the associated linear system (4.21) is easy to solve. But if we consider the diffusion equation in three spatial dimensions, e.g.,

$$u_t = u_{xx} + u_{yy} + u_{zz}, \tag{4.33}$$

   this problem becomes much harder. The solution of linear systems in the form of $Ax = b$, where $A$ is a matrix, $b$ is a known vector, and $x$ is unknown, is a crucial problem in scientific computing. It is a fundamental part of almost all scientific computing projects and presents significant challenges. While small systems are relatively straightforward to solve, larger systems become increasingly difficult. This problem is well studied in the field of scientific computing, but it remains a challenging task in general.

2. *Numerical linear algebra* is a field of research concerned with operations on vectors and matrices on computers. A thorough introduction to the field is given in [1].

3. A classical introduction to iterative methods for solving linear systems is presented in [2].

## References

[1] Lyche T (2020) Numerical Linear Algebra and Matrix Factorizations, vol 22. Springer
[2] Saad Y (2003) Iterative methods for sparse linear systems. SIAM
[3] Tveito A, Winther R (2009) Introduction to partial differential equations; a computational approach, 2nd edn. Springer

# Chapter 5
# Improved Accuracy

In the examples we have considered so far, the error introduced by the numerical scheme has been $O(\Delta t)$. This can easily be improved as we will show through two simple examples.

## 5.1 Back to the Simplest Equation

The first equation we studied was

$$y' = y, \tag{5.1}$$

and we approximated the equation using the finite difference approximation

$$\frac{y_{n+1} - y_n}{\Delta t} = y_n. \tag{5.2}$$

This scheme resulted in an error proportional with the time step $\Delta t$. From Chapter 4, we realize that we can also use the approximation

$$\frac{y_{n+1} - y_n}{\Delta t} = y_{n+1}. \tag{5.3}$$

And, as a compromise between these two alternatives, we can use the following midpoint[1] approximation,

$$\frac{y_{n+1} - y_n}{\Delta t} = \frac{1}{2}(y_n + y_{n+1}). \tag{5.4}$$

The explicit, implicit and midpoint schemes can be written on computational form as follows,

---

[1] This scheme is sometimes called the midpoint scheme (for obvious reasons) and sometimes called the Cranck-Nicolson scheme because it was developed by John Crank and Phyllis Nicolson, [1].

© The Author(s) 2023
K. Horgmo Jæger, A. Tveito, *Differential Equations for Studies in Computational Electrophysiology*,
Simula SpringerBriefs on Computing 14, https://doi.org/10.1007/978-3-031-30852-9_5

$$y_{n+1} = (1 + \Delta t)y_n, \tag{5.5}$$

$$y_{n+1} = \frac{1}{1 - \Delta t} y_n, \tag{5.6}$$

$$y_{n+1} = \frac{1 + \Delta t/2}{1 - \Delta t/2} y_n, \tag{5.7}$$

respectively. In Table 5.1 we show the errors introduced by these three schemes and we note that the midpoint scheme is clearly more accurate than the two other schemes. Furthermore, we find that the error of the implicit and explicit schemes are both proportional to $\Delta t$, or $O(\Delta t)$, whereas the error of the midpoint scheme is proportional to $\Delta x^2$, or $O(\Delta t^2)$. This means that the implicit and explicit schemes have first order (linear) convergence, whereas the midpoint scheme has second order (quadratic) convergence.

**Table 5.1** Errors of numerical solutions of the differential equation (5.1) with initial condition $u_0 = 1$ at $t = T = 1$ for different values of $\Delta t$. The errors are defined as $E_e = |y(1) - y_{N,e}|$, $E_i = |y(1) - y_{N,i}|$ and $E_m = |y(1) - y_{N,m}|$, where $y(1) = e$ is the analytical solution, and $y_{N,e}$, $y_{N,i}$, and $y_{N,m}$, are the numerical solutions of the explicit (5.5), implicit (5.6) and midpoint (5.7) schemes, respectively, at time 1.

| $\Delta t$ | $E_e$ | $E_e/\Delta t$ | $E_i$ | $E_i/\Delta t$ | $E_m$ | $E_m/\Delta t^2$ |
|---|---|---|---|---|---|---|
| 0.1 | 0.125 | 1.25 | 0.15 | 1.5 | 0.00227 | 0.227 |
| 0.01 | 0.0135 | 1.35 | 0.0137 | 1.37 | $2.27 \cdot 10^{-5}$ | 0.227 |
| 0.001 | 0.00136 | 1.36 | 0.00136 | 1.36 | $2.27 \cdot 10^{-7}$ | 0.227 |
| 0.0001 | 0.000136 | 1.36 | 0.000136 | 1.36 | $2.27 \cdot 10^{-9}$ | 0.227 |

The difference in numerical errors reported in Table 5.1 may not seem to be dramatic. But if you require the error to be less than, say, $10^{-10}$, then the number of time steps needed for the first order schemes are about $2.85 \times 10^5$ larger than what is needed for the second order scheme. So, the difference in computing time between a first and second order scheme can be dramatic. For the very simple model considered here, the computation is trivial in any case, but with a challenging system of partial differential equations in three spatial dimensions, the difference in computing efforts can become enormous.

## 5.2 A Linear Reaction-Diffusion Equation

We considered the diffusion equation above. That equation becomes a little more interesting if we add a reaction term to it,

$$u_t = u_{xx} + f(u). \tag{5.8}$$

Here, $f = f(u)$ is referred to as a reaction term. In order to keep things simple, we will consider a linear reaction term and define

$$f(u) = \lambda u, \tag{5.9}$$

and we will wait a little before defining the value of $\lambda$. Above, we noted that a numerical scheme for the diffusion equation could be written in a very compact form by introducing vector/matrix notation. We will extend this in order to define three alternative schemes for the reaction-diffusion equation (5.9). Note that we still apply the following initial and boundary conditions,

$$u(0, x) = u^0(x), \tag{5.10}$$
$$u(t,0) = u(t, 1) = 0. \tag{5.11}$$

As above, we let

$$u^n = \begin{pmatrix} u_2^n \\ u_3^n \\ \vdots \\ u_{M-1}^n \end{pmatrix}, \tag{5.12}$$

and in addition, we introduce the matrix

$$D = \frac{1}{\Delta x^2} \begin{pmatrix} -2 & 1 & 0 & \cdots & & & 0 \\ 1 & -2 & 1 & 0 & \cdots & & 0 \\ 0 & 1 & -2 & 1 & 0 & \cdots & 0 \\ \vdots & & \ddots & \ddots & \ddots & & \vdots \\ 0 & \cdots & & & 0 & 1 & -2 \end{pmatrix}. \tag{5.13}$$

With this notation at hand we can approximate the linear version of equation (5.8) using an explicit, implicit or midpoint approximation of the right-hand side of the equation

$$\frac{u^{n+1} - u^n}{\Delta t} = Du^n + \lambda u^n, \tag{5.14}$$

$$\frac{u^{n+1} - u^n}{\Delta t} = Du^{n+1} + \lambda u^{n+1}, \tag{5.15}$$

$$\frac{u^{n+1} - u^n}{\Delta t} = \frac{1}{2}(Du^n + Du^{n+1} + \lambda u^n + \lambda u^{n+1}). \tag{5.16}$$

To help ease the implementation of the numerical schemes and make it more straightforward to write the schemes in matrix form as in, e.g., (4.21), it is convenient to write these schemes in computational form, that is, collect all terms involving the unknown solutions to be computed in each time step, $u^{n+1}$, on the left-hand side and all the terms involving the know solutions from the previous time step, $u^n$, on the right-hand side. The schemes (5.14)–(5.16) can be written in this form by,

$$u^{n+1} = [(1 + \lambda \Delta t)I + \Delta t D]u^n \qquad (5.17)$$

$$[(1 - \lambda \Delta t)I - \Delta t D]u^{n+1} = u^n \qquad (5.18)$$

$$\left[\left(1 - \frac{\lambda \Delta t}{2}\right)I - \frac{\Delta t}{2}D\right]u^{n+1} = \left[\left(1 + \frac{\lambda \Delta t}{2}\right)I + \frac{\Delta t}{2}D\right]u^n, \qquad (5.19)$$

where $I$ is the identity[2] matrix.

Let us now consider the problem (5.8)-(5.11) with a specific choice of $\lambda = 2\pi^2$ and $u^0(x) = \sin(\pi x)$. Then, the analytical solution is given by

$$u(t, x) = e^{\pi^2 t} \sin(\pi x). \qquad (5.20)$$

In Fig. 5.1, we show the results of the implicit (5.18) and midpoint (5.19) schemes together with the analytical solution at time $T = 1$. In addition, we show the results of the explicit scheme at $T = 0.1$. We have used $\Delta x = 0.01$ and $\Delta t = 0.001$. For this choice of discretization parameters, we get wild oscillations for the explicit scheme. The implicit and midpoint schemes produce more reasonable solutions, and we note that the midpoint scheme is clearly more accurate than the implicit scheme.

## References

[1] Crank J, Nicolson P (1947) A practical method for numerical evaluation of solutions of partial differential equations of the heat-conduction type. In: Mathematical Proceedings of the Cambridge Philosophical Society, Cambridge University Press, vol 43, pp 50–67

---

[2] The identity matrix has ones at the diagonal and zeroes elsewhere.

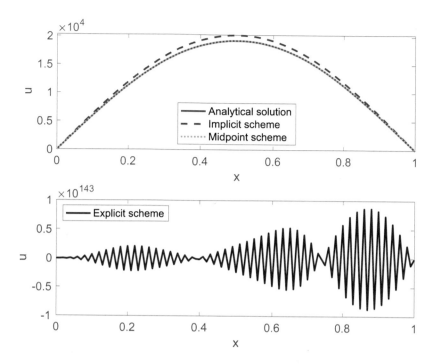

**Fig. 5.1** Analytical and numerical solutions of the reaction-diffusion equation (5.8) with $\lambda = 2\pi^2$, boundary conditions $u(t, 0) = u(t, 1) = 0$ and initial conditions $u(0, x) = \sin(\pi x)$. In the numerical schemes, we use $\Delta t = 0.001$ and $\Delta x = 0.01$. In the upper panel, we show the solution of the implicit and midpoint schemes together with the analytical solution at time $T = 1$. The lower panel shows the solution for the explicit scheme at $T = 0.1$, and we observe wild oscillations.

# Chapter 6
# A Simple Cable Equation

The cable equation was first derived to model transport of electrical signals in telegraphic cables. But it later gained enormous popularity as a model of transport of electrical signals along a neuronal axon. In Chapter 9, we will discuss how this equation is derived and how the different terms in the equation come about. But here, we will just take a simple version of the equations for granted and then try to solve them. We will observe that the few techniques we learned above are actually sufficient to solve the non-linear reaction-diffusion equations we consider here.

## 6.1 A Non-Linear Reaction-Diffusion System

We will consider a system of equations where we add a diffusion term to the FitzHugh-Nagumo equations. Specifically, we consider the equations[1],

$$v_t = \delta v_{xx} + c_1 v(v - a)(1 - v) - c_2 w, \tag{6.1}$$
$$w_t = b(v - dw). \tag{6.2}$$

We use the same constants as above,

$$a = -0.12, \; c_1 = 0.175, \; c_2 = 0.03, \; b = 0.011, \; d = 0.55, \tag{6.3}$$

and in addition we use the diffusion coefficient[2] $\delta = 5 \cdot 10^{-5}$.

---

[1] Recall, once again, that we use the convention that $v_t = \frac{\partial v}{\partial t}$ and $v_{xx} = \frac{\partial^2 v}{\partial x^2}$.

[2] We still use no units; they will be introduced later.

© The Author(s) 2023
K. Horgmo Jæger, A. Tveito, *Differential Equations for Studies in Computational Electrophysiology*,
Simula SpringerBriefs on Computing 14, https://doi.org/10.1007/978-3-031-30852-9_6

## 6.2  The Explicit Numerical Scheme

We note that we can use the same approximation of $v_t$ and $w_t$ that we used for the FitzHugh-Nagumo model in Chapter 2 (see (2.6) and (2.7)), and the same approximation for $v_{xx}$ as we used for the diffusion equation in Chapter 3 (see (3.13)) to define an explicit numerical scheme,

$$\frac{v_j^{n+1} - v_j^n}{\Delta t} = \delta \frac{v_{j-1}^n - 2v_j^n + v_{j+1}^n}{\Delta x^2} + c_1 v_j^n (v_n - a)(1 - v_j^n) - c_2 w_j^n, \tag{6.4}$$

$$\frac{w_j^{n+1} - w_j^n}{\Delta t} = b(v_j^n - dw_j^n). \tag{6.5}$$

Here, $v_j^n$ and $w_j^n$ denote approximations of $v(x_j, t_n)$ and $w(x_j, t_n)$, respectively. It is straightforward to put this scheme in computational form,

$$v_j^{n+1} = \rho v_{j-1}^n + (1 - 2\rho)v_j^n + \rho v_{j+1}^n + \Delta t[c_1 v_j^n (v_n - a)(1 - v_j^n) - c_2 w_j^n],$$
$$w_j^{n+1} = w_j^n + \Delta t b(v_j^n - dw_j^n)$$

where $\rho = \delta \Delta t / \Delta x^2$.

## 6.3  Traveling Wave Solutions

Traveling wave solutions are characteristic of solutions of reaction-diffusion models of neuronal axons and myocardial tissue. Here, we will see such solutions for the simple model given by (6.1) and (6.2).

In Fig. 6.1, we show the numerical solution of $v$ as a function of $x$ at five different points in time. In order to initiate a wave traveling from left to right, we let the initial conditions be specified by $v_0 = 0$ and $w_0 = 0$ for all values of $x$, except that we set $v_0 = 0.26$ for $x \leq 0.04$. We use $\Delta x = 0.01$ and $\Delta t = 0.005$. The boundary conditions for $v$ are given by $\frac{\partial v}{\partial x}(t, 0) = \frac{\partial v}{\partial x}(t, 1) = 0$.

In the leftmost panel of Fig. 6.1, at $t = 0$, we see the initial conditions for the variable $v$. In the second panel, at $t = 50$, we see that the value of $v$ has started to increase in an area close to the left boundary of the domain, and in the next panels, at $t = 100$, $t = 200$, and $t = 300$, we see that the increase in $v$ gradually moves from left to right like a traveling wave.

### 6.3.1  Adjusting Parameters

In Fig. 6.2, we show the solution of the model at five different points in time for three different values of the parameters $\delta$. We observe that for a low value of $\delta$

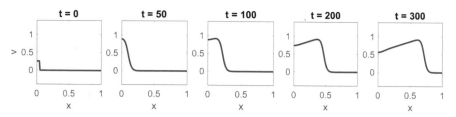

**Fig. 6.1** Numerical solution of $v$ in the FitzHugh-Nagumo equations with a diffusion term added. The boundary conditions are given by $\frac{\partial v}{\partial x}(t, 0) = \frac{\partial v}{\partial x}(t, 1) = 0$, and the initial conditions are $w(0, x) = 0$ everywhere and $v(0, x) = 0$ for $x > 0.04$ and $v(0, x) = 0.26$ for $x \leq 0.04$. We use $\Delta t = 0.005$ and $\Delta x = 0.01$, and show the solution at five different points in time.

($\delta = 1 \cdot 10^{-5}$), the traveling wave moves more slowly through the domain. For example, at $t = 200$, the wave has crossed about half of the distance from $x = 0$ to $x = 1$ for the default case of $\delta = 5 \cdot 10^{-5}$, and only about a quarter of the distance for $\delta = 1 \cdot 10^{-5}$. Moreover, for a high value of $\delta$ ($\delta = 20 \cdot 10^{-5}$), the wave moves more quickly through the domain, and has almost crossed the entire domain at $t = 200$. We also observe that the slope of the wavefront of the traveling wave appears to become less steep at the value of $\delta$ is increased.

In Fig. 6.3, we similarly consider the traveling wave solutions for three different values of the parameter $c_1$. As when we adjust the value of $\delta$, we observe that the wave moves more quickly as the value of $c_1$ is increased.

### 6.3.2 Conduction Velocity

In Fig. 6.2 and Fig. 6.3, we observed that the traveling wave moves more quickly as we increased the value of $\delta$ or $c_1$. The conduction velocity is often used to characterize the speed with which a traveling wave traverses the domain. In Fig. 6.3, we have computed the conduction velocity for some different values of $\delta$ and $c_1$. Here, we have defined the conduction velocity as

$$CV = \frac{x_2 - x_1}{t_2 - t_1}, \tag{6.6}$$

where $x_1 = 0.5$ and $x_2 = 0.7$. Furthermore, $t_1$ and $t_2$ are the points in time when the value of $v$ first increases to a value $v \geq 0.5$, in the spatial points $x_1$ and $x_2$, respectively. As expected Fig. 6.3 shows that increasing $\delta$ or $c_1$ in the model increases the computed conduction velocity.

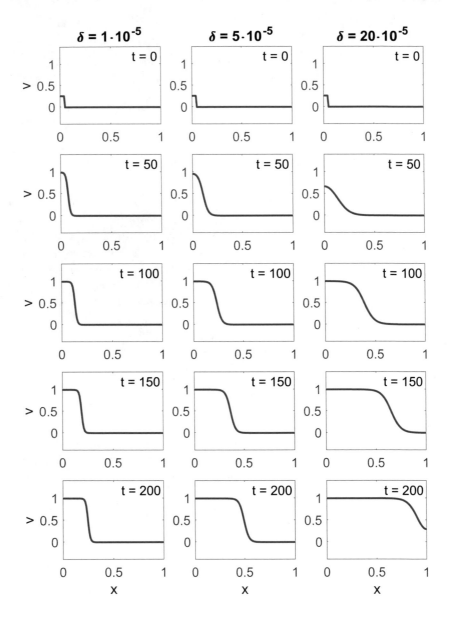

**Fig. 6.2** Numerical solution of $v$ in the FitzHugh-Nagumo equations with an added diffusion term at for three different values of $\delta$. The remaining parameter values are as specified in (6.3). The boundary conditions are given by $\frac{\partial v}{\partial x}(t, 0) = \frac{\partial v}{\partial x}(t, 1) = 0$, and the initial conditions are $w(0, x) = 0$ everywhere and $v(0, x) = 0$ for $x > 0.04$ and $v(0, x) = 0.26$ for $x \leq 0.04$. We use $\Delta t = 0.005$ and $\Delta x = 0.01$, and show the solution at five different points in time.

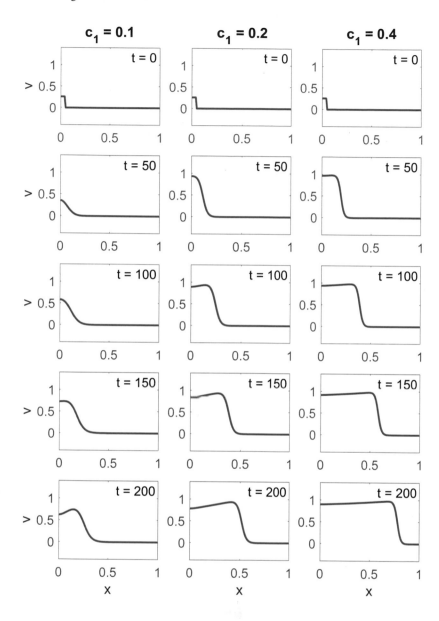

**Fig. 6.3** Numerical solution of $v$ in the FitzHugh-Nagumo equations with an added diffusion term for three different values of $\delta$ for three different values of $c_1$. The remaining parameter values are as specified in (6.3) and $\delta = 5 \cdot 10^{-5}$. The boundary conditions are given by $\frac{\partial v}{\partial x}(t, 0) = \frac{\partial v}{\partial x}(t, 1) = 0$, and the initial conditions are $w(0, x) = 0$ everywhere and $v(0, x) = 0$ for $x > 0.04$ and $v(0, x) = 0.26$ for $x \le 0.04$. We use $\Delta t = 0.005$ and $\Delta x = 0.01$, and show the solution at five different points in time.

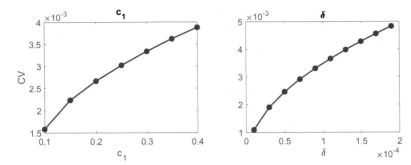

**Fig. 6.4** Conduction velocity for different values of $\delta$ and $c_1$, computed from numerical solutions of the FitzHugh-Nagumo equations with an added diffusion equation term. The conduction velocity is computed as defined in (6.6).

# Chapter 7
# Operator Splitting

In mathematics, a common approach to solving a new problem is to break down the problem into sub-problems that you know how to solve and then try to glue the pieces together to yield a solution of the new problem. This approach is also very useful in software development; well-tested pieces of software can be glued together in order to obtain solutions to a wider class of problems. Operator splitting is a technique that illustrates this principle very well. Rather complex equations can be broken down into more familiar problems and solved individually. It's a miracle that it works, but it does. And it's no miracle because there are proofs of convergence. Anyway, we will illustrate operator splitting with two examples and then come back to this technique when the equations become more challenging.

## 7.1 Numerical Schemes for a Reaction-Diffusion Equation

Suppose we want to solve the following reaction-diffusion equation,

$$u_t = u_{xx} + f(u), \tag{7.1}$$

with initial and boundary conditions,

$$u(0, x) = u^0(x), \tag{7.2}$$
$$u(t, 0) = u(t, 1) = 0. \tag{7.3}$$

Clearly, we can use the techniques introduced above. By using the notation introduced on page 43, we can write explicit, implicit, and midpoint schemes as follows,

© The Author(s) 2023
K. Horgmo Jæger, A. Tveito, *Differential Equations for Studies in Computational Electrophysiology*, Simula SpringerBriefs on Computing 14, https://doi.org/10.1007/978-3-031-30852-9_7

$$\frac{u^{n+1} - u^n}{\Delta t} = Du^n + f(u^n), \tag{7.4}$$

$$\frac{u^{n+1} - u^n}{\Delta t} = Du^{n+1} + f(u^{n+1}), \tag{7.5}$$

$$\frac{u^{n+1} - u^n}{\Delta t} = \frac{1}{2}(Du^n + Du^{n+1}) + \frac{1}{2}(f(u^n) + f(u^{n+1})), \tag{7.6}$$

where we have tacitly extended $f$ to be a vector valued function with components $f_i = f(u_i)$. These schemes can be rearranged to computational form as follows,

$$u^{n+1} = [I + \Delta t D]u^n + \Delta t f(u^n),$$

$$[I - \Delta t D]u^{n+1} - \Delta t f(u^{n+1}) = u^n,$$

$$\left(I - \frac{\Delta t}{2}D\right)u^{n+1} - \frac{\Delta t}{2}f(u^{n+1}) = \left(I + \frac{\Delta t}{2}D\right)u^n + \frac{\Delta t}{2}f(u^n),$$

where, again, we use the convention that known quantities are on the right-hand side of the equations and the unknowns are at the left-hand side. The explicit scheme is straightforward to implement since computing $u^{n+1}$ simply amounts to evaluating the right-hand side. But the implicit scheme has become more complicated than we are used to because we have to solve a potentially non-linear *system of algebraic equations* in order to compute $u^{n+1}$. We can do that using Newton's method, but we can also avoid this by introduction operator splitting.

## 7.2 Operator Splitting for a Reaction-Diffusion Equation

Let $u^n$ denote an approximation of $u(t_n, x)$, where $t_n = n\Delta t$ as usual. Then, the problem (7.1) can be solved by alternately solving $u_t = u_{xx}$ and $u_t = f(u)$. More precisely, we assume that an approximate solution has been computed for time $t = t_n$. Then, an approximate solution of (7.1) at $t_{n+1}$ can be found in two steps. First we solve

$$u_t = u_{xx}, \tag{7.7}$$

from $t_n$ to $t_{n+1}$ with the initial condition $u(t_n, \cdot) = u^n$ and boundary conditions given by (7.2) and (7.3). We let $u^{n+1/2}$ denote the solution of the first step. In the second step, we solve

$$u_t = f(u), \tag{7.8}$$

using the initial condition $u(t_n, \cdot) = u^{n+1/2}$. Now, we have broken the somewhat complex problem (7.1) down to two problems we are more familiar with. We will show how this works with a couple of examples. Note, however, that the main message from this chapter is the technique of breaking down the numerical solution of a problem into two simpler problems like described in this subsection. So if you reach a point where you feel that the mathematics in the remaining part of this chapter

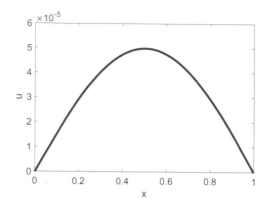

**Fig. 7.1** Numerical solution of the differential equation $u_t = u_{xx} - u^2$, with boundary conditions $u(t, 0) = u(t, 1) = 0$ and initial condition $u(0, x) = \sin(\pi x)$. The problem is solved using the operator splitting procedure described in (7.10)–(7.13) with $\Delta x = 0.01$ and $\Delta t = 0.001$.

becomes a bit overwhelming, it might be good to jump ahead to Chapter 8, where we will start applying the methods introduced above to models of electrophysiology.

### 7.2.1 Numerical Example

In the first numerical example, we consider the problem

$$u_t = u_{xx} - u^2, \tag{7.9}$$

with the boundary conditions $u(t,0) = u(t,1) = 0$ and the initial condition $u(0, x) = \sin(\pi x)$. We want to use the operator splitting procedure introduced above to solve this problem numerically, and this will result in two steps. The first step is to solve

$$u_t = u_{xx}, \tag{7.10}$$

with an implicit scheme (see page 34). The second step is to solve

$$u_t = -u^2, \tag{7.11}$$

using an implicit scheme. Note that this equation has to be solved for each mesh point $x_i$, so the implicit scheme reads

$$\frac{u_j^{n+1} - u_j^{n+1/2}}{\Delta t} = -(u_j^{n+1})^2, \tag{7.12}$$

where $u_j^{n+1/2}$ is the result of the first step. By using operator splitting we have avoided solving a big system of non-linear equations. Instead we need to solve a usual linear system of equations arising from the discrete version of (7.10) and a series of second order algebraic equations given by (7.12) whose solutions are given by the quadratic formula,

$$u_j^{n+1} = \frac{-1 + \sqrt{1 + 4\Delta t u_j^{n+1/2}}}{2\Delta t}.$$  (7.13)

In Fig. 7.1 we show the solution of this problem using $\Delta x = 0.01$ and $\Delta t = 0.001$. Furthermore, in Table 7.1 we have computed the error by comparing the solution to a very fine scale solution of this problem using a standard explicit scheme. Again we note that the error seems to be first order in $\Delta t$.

Table 7.1 Errors of the numerical solutions of the differential equation $u_t = u_{xx} - u^2$, with boundary conditions $u(t, 0) = u(t, 1) = 0$ and initial condition $u(0, x) = \sin(\pi x)$ for different values of $\Delta t$. The error is defined as $E = \max_j |u_{e,j} - u_j^N|$, where $u_{e,j}$ is the solution of the problem found using a standard explicit scheme with a very fine time step ($\Delta t = 10^{-6}$), and $u_j^N$ is the numerical solution of the operator splitting scheme described in (7.10)–(7.13). We use $\Delta x = 0.01$.

| $\Delta t$ | $E$ | $E/\Delta t$ |
|---|---|---|
| 0.01 | $2.77 \cdot 10^{-5}$ | 0.0028 |
| 0.005 | $1.27 \cdot 10^{-5}$ | 0.0025 |
| 0.001 | $2.37 \cdot 10^{-6}$ | 0.0024 |
| 0.0005 | $1.17 \cdot 10^{-6}$ | 0.0023 |
| 0.0001 | $2.35 \cdot 10^{-7}$ | 0.0024 |

## 7.2.2 A Detour via Numerical Integration

Very few things come for free, but second order convergence does. With minimal change of the algorithm above, we can obtain second order convergence. Actually, just the first and the last step of the algorithm are slightly changed (half step instead of whole step). Exactly the same alteration is present in changing from a first order to a second order scheme of numerical integration. We will show that similarity here because it might help make the second order operator splitting algorithm seem less mysterious, but if you are not interested, just skip this subsection and go to page 58 where you can read about second order operator splitting.

### First and Second Order Schemes for Numerical Integration

Suppose we have a smooth function $g = g(x)$ defined on the interval from 0 to 1, and that we want to compute an approximation of the integral of $g$ on this interval. We start by defining $\Delta x = 1/(M - 1)$ and $x_i = (i - 1) \times \Delta x$ for a sufficiently large integer $M$. In Fig. 7.2, we show two approximations of the function $g$. The first

**Piecewise constant approximation**

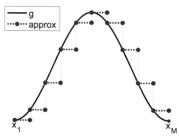

**Piecewise linear approximation**

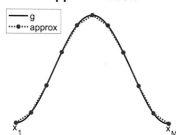

**Fig. 7.2** Illustration of two approximations of a function $g$. In the left panel, we illustrate a piecewise constant approximation defined by the value of $g$ at $x = x_i$ for the interval between $x_i$ to $x_{i+1}$. In the right panel, we illustrate a piecewise linear approximation defined to coincide with $g$ for $x = x_i$ and $x = x_{i+1}$.

approximation is simply a constant function defined by the value of $g$ at $x = x_i$. The second is a linear function that coincides with $g$ for $x = x_i$ and $x = x_{i+1}$.

If we use these two approximations to estimate the integral of $g$ from $x = x_i$ to $x = x_{i+1}$, we get

$$\int_{x_i}^{x_{i+1}} g(x)dx \approx \Delta x g(x_i) \tag{7.14}$$

and

$$\int_{x_i}^{x_{i+1}} g(x)dx \approx \frac{1}{2}\Delta x(g(x_i) + g(x_{i+1})), \tag{7.15}$$

respectively. Since we clearly have

$$\int_0^1 g(x)dx = \sum_{i=1}^{M-1} \int_{x_i}^{x_{i+1}} g(x)dx, \tag{7.16}$$

we get two approximations of the integral from (7.15) and (7.16), respectively;

$$\int_0^1 g(x)dx \approx \Delta x \sum_{i=1}^{M-1} g(x_i), \tag{7.17}$$

and

$$\int_0^1 g(x)dx \approx \frac{1}{2}\Delta x \sum_{i=1}^{M-1} (g(x_i) + g(x_{i+1})). \tag{7.18}$$

Here, the latter formula can be rewritten slightly to

$$\int_0^1 g(x)dx \approx \Delta x \left[ \frac{1}{2}g(x_1) + g(x_2) + g(x_3) + \cdots + g(x_{M-1}) + \frac{1}{2}g(x_M) \right]. \tag{7.19}$$

**Table 7.2** Maximum errors of the Riemann sum approximation ($E_R$, (7.17)) and the Trapezoidal scheme approximation ($E_T$, (7.19)) to the integral $\int_0^1 x^2 dx$ for some different values of $\Delta x$.

| $\Delta x$ | $E_R$ | $E_R/\Delta x$ | $E_T$ | $E_T/\Delta x^2$ |
|---|---|---|---|---|
| 0.1 | 0.0483 | 0.48 | 0.00167 | 0.17 |
| 0.02 | 0.00993 | 0.5 | $6.67 \cdot 10^{-5}$ | 0.17 |
| 0.01 | 0.00498 | 0.5 | $1.67 \cdot 10^{-5}$ | 0.17 |
| 0.002 | 0.000999 | 0.5 | $6.67 \cdot 10^{-7}$ | 0.17 |
| 0.001 | 0.0005 | 0.5 | $1.67 \cdot 10^{-7}$ | 0.17 |

The approximation of the integral defined by (7.17) is referred to as a Riemann sum, whereas the approximation given by (7.19) is referred to as the Trapezoidal method of integration. In Table 7.2, we show the error when using these to schemes to approximate the integral

$$\int_0^1 x^2 dx = \frac{1}{3}$$

using several values of $\Delta x$. For the Riemann sum, the error is $O(\Delta x)$, and for the Trapezoidal scheme, the error is $O(\Delta x^2)$. So, by using the same number of function evaluations, the accuracy of the integration scheme is improved from first to second order.

### 7.2.3 Second Order Operator Splitting

In order to introduce second order operator splitting, it is useful to introduce notations that we can use to simplify the formulations. We recall that the problem we want to solve is the following reaction-diffusion equation,

$$u_t = u_{xx} + f(u), \tag{7.20}$$

with initial and boundary conditions,

$$u(0, x) = u^0(x), \tag{7.21}$$
$$u(t, 0) = u(t, 1) = 0. \tag{7.22}$$

First, we let $\mathcal{D}(\Delta t)$ be an operator that evolves the solution of the diffusion equation one time step $\Delta t$ ahead. If the function $u = u(t, \cdot)$ is known at time $t$, then $u(t + \Delta t, \cdot) = \mathcal{D}(\Delta t)u(t, \cdot)$ denotes the solution of

$$u_t = u_{xx}, \tag{7.23}$$

at time $t + \Delta t$, where $u(t)$ is the initial condition at time $t$. Similarly, we let $\mathcal{R}$ play the same role for the reaction part of the equation. If $u = u(t, \cdot)$ is known, then $u(t + \Delta t, \cdot) = \mathcal{R}(\Delta t)u(t, \cdot)$ denotes the solution of

$$u_t = f(u), \tag{7.24}$$

at time $t + \Delta t$, where $u(t)$ is the initial condition at time $t$. With this notation, we can rewrite the first order operator splitting derived above in a very compact manner. The step from $t_n$ to $t_{n+1}$ can be written

$$u^{n+1} = \mathcal{R}(\Delta t)\mathcal{D}(\Delta t)u^n. \tag{7.25}$$

By simply repeating the process, we find that

$$u^n = (\mathcal{R}(\Delta t)\mathcal{D}(\Delta t))^n u^0. \tag{7.26}$$

Equipped with this notation, we can improve the accuracy of the operator splitting using the same approach as for numerical integration (7.19). In order to improve the accuracy of the operator splitting to second order, we now approximate one time step by

$$u^{n+1} = \mathcal{D}(\Delta t/2)\mathcal{R}(\Delta t)\mathcal{D}(\Delta t/2)u^n. \tag{7.27}$$

By combining several steps, we get

$$u^n = [\mathcal{D}(\Delta t/2)\mathcal{R}(\Delta t)\mathcal{D}(\Delta t/2) \cdots \mathcal{D}(\Delta t/2)\mathcal{R}(\Delta t)\mathcal{D}(\Delta t/2)]u^0. \tag{7.28}$$

Note that by the definition of $\mathcal{D}$, it has the following useful property,

$$\mathcal{D}(\Delta t/2)\mathcal{D}(\Delta t/2) = \mathcal{D}(\Delta t),$$

since applying $\mathcal{D}(\Delta t/2)$ twice simply means to solve the equation twice using the time step $\Delta t/2$. By using this property, we can rewrite (7.28) as follows,

$$u^n = \left( \mathcal{D}(\Delta t/2)[\mathcal{R}(\Delta t)\mathcal{D}(\Delta t)]^{n-1}\mathcal{R}(\Delta t)\mathcal{D}(\Delta t/2) \right) u^0. \tag{7.29}$$

So the only difference between the first order scheme (7.26) and the second order scheme (7.29) is that in the latter, we do a half time step in the first and last time steps. This is very similar to the half step in the first and last step of the numerical integration above; see (7.19) compared to (7.17).

## 7.3 Operator Splitting Applied to a System of Reaction-Diffusion Equations

In order to demonstrate the use of the operator splitting techniques introduced above, we revisit the FitzHugh-Nagumo system,

$$v_t = \delta v_{xx} + c_1 v(v - a)(1 - v) - c_2 w, \tag{7.30}$$

$$w_t = b(v - dw). \tag{7.31}$$

We use the same constants as above,

$$a = -0.12, \ c_1 = 0.175, \ c_2 = 0.03, \ b = 0.011, \ d = 0.55, \ \delta = 5 \cdot 10^{-5}. \tag{7.32}$$

Here, the solution operator of the diffusion step $\mathcal{D}$ evolves

$$v_t = \delta v_{xx}, \tag{7.33}$$

$$w_t = 0, \tag{7.34}$$

and the $\mathcal{R}$ evolves

$$v_t = c_1 v(v - a)(1 - v) - c_2 w, \tag{7.35}$$

$$w_t = b(v - dw). \tag{7.36}$$

In Table 7.3 we show the error when the first order (see (7.26)) and second order (see (7.29)) algorithms. The solutions are computed by replacing $\mathcal{D}$ and $\mathcal{R}$ by the standard explicit schemes with a fine time step ($\Delta t = 10^{-4}$), and the errors are estimated by comparing the solutions to a numerical solution found using the fine time step and no operator splitting. As anticipated, the convergence of the two methods are first and second order.

**Table 7.3** Maximum errors of the first order ($E_1$, (7.26)) and the second order ($E_2$, (7.29)) operator splitting techniques applied to the FitzHugh-Nagumo system (7.30)–(7.32). In each operator splitting step, the system is solved using standard explicit schemes with $\Delta t = 10^{-4}$ and $\Delta x = 0.01$. The error is found by comparing the solutions to solutions found using a standard explicit scheme with $\Delta t = 10^{-4}$ and $\Delta x = 0.01$ without operator splitting.

| $\Delta t$ | $E_1$ | $E_1/\Delta t$ | $E_2$ | $E_2/\Delta t^2$ |
|---|---|---|---|---|
| 5 | 0.0205 | 0.0041 | 0.00612 | 0.00024 |
| 2 | 0.00768 | 0.0038 | 0.0011 | 0.00028 |
| 1 | 0.00384 | 0.0038 | 0.000296 | 0.00030 |
| 0.5 | 0.00192 | 0.0038 | $7.48 \cdot 10^{-5}$ | 0.00030 |
| 0.2 | 0.000765 | 0.0038 | $1.05 \cdot 10^{-5}$ | 0.00026 |

## 7.4 Comments and Further Reading

1. The example in Section 7.3 indicates the strength of operator splitting. If we have a good solver for the diffusion step, and a good solver for a system of

ordinary differential equations, these solvers can be combined to get a solution of the reaction diffusion problem. In this case, this is not really necessary, but there are extremely complex problems out there that are virtually impossible to solve without using operator splitting.

2. Why does operator splitting work? This question is studied in some detailed in, e.g., [4, 5, 7, 8, 9, 12]. A detailed discussion of this topic is far outside of our scope. However, we can give a hint to why this works. To this end, we consider a linear system of ordinary differential equations

$$u_t = Au + Bu,$$

where $u$ is a vector and $A$ and $B$ are matrices compatible with the dimensions of $u$; they may represent discrete version of spatial derivatives as above. An explicit scheme for this system can be written on the form

$$u^{n+1} = (I + \Delta t A + \Delta t B)u^n,$$

and using first order operator splitting, we get the scheme

$$U^{n+1/2} = (I + \Delta t A)U^n, \tag{7.37}$$

$$U^{n+1} = (I + \Delta t B)U^{n+1/2}. \tag{7.38}$$

By combining the two steps of the operator splitting scheme, we get

$$U^{n+1} = (I + \Delta t A)(I + \Delta t B)U^n. \tag{7.39}$$

This scheme can be expanded to read

$$U^{n+1} = (I + \Delta t A + \Delta t B)U^n + \Delta t^2 ABU^n. \tag{7.40}$$

Hence, in the step from $t_n$ to $t_{n+1}$ the results of the two schemes differ by $O(\Delta t^2)$, and thus the difference summed over $N \sim \Delta t^{-1}$ steps is $O(\Delta t)$. Therefore, the standard explicit scheme and the first order operator splitting scheme converges to the same solution as $\Delta t$ goes to zero.

3. The first order operator splitting is referred to as Godunov splitting [3] and second order splitting is referred to as Strang splitting [10]. In [2], these methods are used to split spatial parts of the differential equation. In that setting, the method is referred to as the method of fractional steps.

4. In computational electrophysiology, operator splitting was introduced for the monodomain model in [6] and extended to the bidomain model in [11]. The error of the method was analyzed in [7], and higher order splitting methods were introduced in [1].

5. In this chapter, we have learned that operator splitting is an effective method for breaking down complex problems into simpler ones. This technique is widely used in mathematics and is particularly useful in computational mathematics, as it enables the reuse of code. However, the question remains whether a

coupled system is inherently more difficult to solve. It is possible to solve a reaction-diffusion equation using a fully coupled implicit scheme and Newton's method, but when the problem involves large systems of equations defined on different domains and scales, and it can be impractical to solve the problem in a fully coupled manner. In such cases, operator splitting is often necessary to manage the complexity of the problem.

# References

[1] Cervi J, Spiteri RJ (2018) High-order operator splitting for the bidomain and monodomain models. SIAM Journal on Scientific Computing 40(2):A769–A786

[2] Crandall M, Majda A (1980) The method of fractional steps for conservation laws. Numerische Mathematik 34(3):285–314

[3] Godunov S, Bohachevsky I (1959) Finite difference method for numerical computation of discontinuous solutions of the equations of fluid dynamics. Matematičeskij Sbornik 47(3):271–306

[4] Karlsen KH, Risebro NH (1997) An operator splitting method for nonlinear convection-diffusion equations. Numerische Mathematik 77(3):365–382

[5] LeVeque R (2002) Finite Volume Methods for Hyperbolic Problems. Cambridge Texts in Applied Mathematics

[6] Qu Z, Garfinkel A (1999) An advanced algorithm for solving partial differential equation in cardiac conduction. IEEE Transactions on Biomedical Engineering 46(9):1166–1168

[7] Schroll HJ, Lines GT, Tveito A (2007) On the accuracy of operator splitting for the monodomain model of electrophysiology. International Journal of Computer Mathematics 84(6):871–885

[8] Spiteri RJ, Ziaratgahi ST (2016) Operator splitting for the bidomain model revisited. Journal of Computational and Applied Mathematics 296:550–563

[9] Sportisse B (2000) An analysis of operator splitting techniques in the stiff case. Journal of computational physics 161(1):140–168

[10] Strang G (1968) On the construction and comparison of difference schemes. SIAM Journal on Numerical Analysis 5(3):506–517

[11] Sundnes J, Lines GT, Tveito A (2005) An operator splitting method for solving the bidomain equations coupled to a volume conductor model for the torso. Mathematical Biosciences 194(2):233–248

[12] Sundnes J, Lines GT, Cai X, Nielsen BF, Mardal KA, Tveito A (2006) Computing the electrical activity in the heart. Springer

# Part II
# Models of Electrophysiology

# Chapter 8
# Membrane Models

In the previous chapters, we introduced techniques that can be used to find numerical solutions of differential equations. The examples we considered were simple, theoretical differential equations without units. From this point forwards, we will look at how to apply the methods introduced in the previous chapters to differential equations that are set up to model aspects of electrophysiology.

In this chapter, we will consider a type of model that is commonly used to model the dynamics across the membrane of excitable cells. We will start by introducing a model for the action potentials generated in neurons. More specifically, we will consider the Hodgkin-Huxley model for the action potential of the squid giant axon [10]. Then, we will introduce a similar model for a cardiac action potential — specifically, a model for the action potential of a rabbit ventricular cardiomyocyte [9].

## 8.1 The Hodgkin-Huxley Model

The Hodgkin-Huxley model [10] consists of a system of four ordinary differential equations with the four unknowns $v$, $m$, $h$, and $r$[1]:

$$C_m \frac{dv}{dt} = -(I_{Na} + I_K + I_L), \tag{8.1}$$

$$\frac{dm}{dt} = \alpha_m(1 - m) - \beta_m m, \tag{8.2}$$

$$\frac{dh}{dt} = \alpha_h(1 - h) - \beta_h h, \tag{8.3}$$

$$\frac{dr}{dt} = \alpha_r(1 - r) - \beta_r r. \tag{8.4}$$

---

[1] The unknown function $r$ is actually called $n$ in the original formulation of the Hodgkin-Huxley model, but we will use the name $r$ to avoid confusion with the index $n$ used to denote the time step in the numerical schemes.

© The Author(s) 2023
K. Horgmo Jæger, A. Tveito, *Differential Equations for Studies in Computational Electrophysiology*, Simula SpringerBriefs on Computing 14, https://doi.org/10.1007/978-3-031-30852-9_8

Here, the unknown function $v$ has unit millivolts (mV) and represents the membrane potential, and $C_m = 1\ \mu\text{F/cm}^2$ is a parameter representing the specific membrane capacitance. Furthermore, $I_{\text{Na}}$, $I_{\text{K}}$, and $I_{\text{L}}$ represent the current density through three types of ion channel: Na$^+$ channels, K$^+$ channels, and non-specific leak channels. The current densities are given in units of $\mu\text{A/cm}^2$ and are defined by

$$I_{\text{Na}} = g_{\text{Na}} m^3 h(v - v_{\text{Na}}), \tag{8.5}$$

$$I_{\text{K}} = g_{\text{K}} r^4(v - v_{\text{K}}), \tag{8.6}$$

$$I_{\text{L}} = g_{\text{L}}(v - v_{\text{L}}), \tag{8.7}$$

where $g_{\text{Na}} = 120$ mS/cm$^2$, $g_{\text{K}} = 36$ mS/cm$^2$ and $g_{\text{L}} = 0.3$ mS/cm$^2$ are parameters representing the maximal conductance density of the different channel types, and $v_{\text{Na}} = 50$ mV, $v_{\text{K}} = -77$ mV and $v_{\text{L}} = -54.4$ mV are the equilibrium potentials of the channels. In addition, $m^3 h$ and $r^4$ represent the open probability of the Na$^+$ and the K$^+$ channels, respectively. The open probability of the leak channels is assumed to be 1 at all times.

The unknown functions $m$, $h$ and $r$ take values between 0 and 1 and are governed by the equations (8.2)–(8.4). In these equations, $\alpha_m, \beta_m, \alpha_h, \beta_h, \alpha_r$, and $\beta_r$ represent rates of the opening and closing of channel gates, are given in units of ms$^{-1}$ and depend on the value of $v$. More specifically, they are defined by

$$\alpha_m = \frac{\gamma_1(v + \gamma_2)}{1 - e^{-(v+\gamma_2)/\gamma_3}}, \qquad\qquad \beta_m = \gamma_4 e^{-(v+\gamma_5)/\gamma_6}, \tag{8.8}$$

$$\alpha_h = \gamma_7 e^{-(v+\gamma_8)/\gamma_9}, \qquad\qquad \beta_h = \frac{\gamma_{10}}{1 + e^{-(v+\gamma_{11})/\gamma_{12}}}, \tag{8.9}$$

$$\alpha_r = \frac{\gamma_{13}(v + \gamma_{14})}{1 - e^{-(v+\gamma_{14})/\gamma_{15}}}, \qquad\qquad \beta_r = \gamma_{16} e^{-(v+\gamma_{17})/\gamma_{18}}, \tag{8.10}$$

where the parameters $\gamma_1$–$\gamma_{18}$ are constants specified by

$$\gamma_1 = 0.1\ \text{ms}^{-1}\text{mV}^{-1}, \qquad \gamma_2 = 40\ \text{mV}, \qquad \gamma_3 = 10\ \text{mV}, \tag{8.11}$$

$$\gamma_4 = 4\ \text{ms}^{-1}, \qquad \gamma_5 = 65\ \text{mV}, \qquad \gamma_6 = 18\ \text{mV}, \tag{8.12}$$

$$\gamma_7 = 0.07\ \text{ms}^{-1}, \qquad \gamma_8 = 65\ \text{mV}, \qquad \gamma_9 = 20\ \text{mV}, \tag{8.13}$$

$$\gamma_{10} = 1\ \text{ms}^{-1}, \qquad \gamma_{11} = 35\ \text{mV}, \qquad \gamma_{12} = 10\ \text{mV}, \tag{8.14}$$

$$\gamma_{13} = 0.01\ \text{ms}^{-1}\text{mV}^{-1}, \qquad \gamma_{14} = 55\ \text{mV}, \qquad \gamma_{15} = 10\ \text{mV}, \tag{8.15}$$

$$\gamma_{16} = 0.125\ \text{ms}^{-1}, \qquad \gamma_{17} = 65\ \text{mV}, \qquad \gamma_{18} = 80\ \text{mV}. \tag{8.16}$$

In our computations reported below, we will use the initial conditions

$$v(0) = -60\ \text{mV}, \qquad m(0) = 0.1, \qquad h(0) = 0.6, \qquad r(0) = 0.3. \tag{8.17}$$

### 8.1.1 An Explicit Numerical Scheme

The differential equations of the Hodgkin-Huxley model can be solved numerically using the techniques we have applied for simple example equations in the previous chapters. More specifically, we can define a numerical scheme for the equations by replacing the derivatives in (8.1)–(8.4) by the standard difference,

$$f'(t) \approx \frac{f(t + \Delta t) - f(t)}{\Delta t}, \tag{8.18}$$

and define an explicit scheme by

$$C_m \frac{v_{n+1} - v_n}{\Delta t} = -(I_{Na}(v_n, m_n, h_n) + I_K(v_n, r_n) + I_L(v_n)), \tag{8.19}$$

$$\frac{m_{n+1} - m_n}{\Delta t} = \alpha_m(v_n)(1 - m_n) - \beta_m(v_n)m_n, \tag{8.20}$$

$$\frac{h_{n+1} - h_n}{\Delta t} = \alpha_h(v_n)(1 - h_n) - \beta_h(v_n)h_n, \tag{8.21}$$

$$\frac{r_{n+1} - r_n}{\Delta t} = \alpha_r(v_n)(1 - r_n) - \beta_r(v_n)r_n, \tag{8.22}$$

where, as usual, $v_n$, $m_n$, $h_n$, and $r_n$ are the numerical solutions at time $t_n = n \times \Delta t$. The scheme can be written in computational form as

$$v_{n+1} = v_n - \frac{\Delta t}{C_m}[I_{Na}(v_n, m_n, h_n) + I_K(v_n, r_n) + I_L(v_n)], \tag{8.23}$$

$$m_{n+1} = m_n + \Delta t[\alpha_m(v_n)(1 - m_n) - \beta_m(v_n)m_n], \tag{8.24}$$

$$h_{n+1} = h_n + \Delta t[\alpha_h(v_n)(1 - h_n) - \beta_h(v_n)h_n], \tag{8.25}$$

$$r_{n+1} = r_n + \Delta t[\alpha_r(v_n)(1 - r_n) - \beta_r(v_n)r_n]. \tag{8.26}$$

### 8.1.2 Error of the Numerical Solution

In Table 8.1, we report the error, $E_v$, of the numerical solution of $v$ using the numerical scheme (8.23)–(8.26) for some different values of $\Delta t$. The error is computed by comparing the solutions at time $t = 3$ ms to the solutions found using a very fine time step ($\Delta t = 10^{-6}$ ms). As observed for simpler systems of equations in earlier chapters, we find that the error of the explicit scheme is close to proportional to the time step, $\Delta t$, applied in the numerical scheme. In other words, we have linear (or first order) convergence.

**Table 8.1** Error of the numerical solution of the Hodgkin-Huxley model for different values of $\Delta t$. The error is defined as $E_v = |v - v_N|$, where $v$ is the numerical solution at $t = 3$ ms for a very fine time step ($\Delta t = 10^{-6}$ ms), and $v_N$ is the numerical solution at $t = 3$ ms for each of the values of $\Delta t$ reported in the first column of the table.

| $\Delta t$ (ms) | $E_v$ (mV) | $E_v/\Delta t$ (mV/ms) |
| --- | --- | --- |
| 0.01 | 0.982 | 98 |
| 0.005 | 0.49 | 98 |
| 0.001 | 0.0979 | 98 |
| 0.0005 | 0.0489 | 98 |
| 0.0001 | 0.0097 | 97 |

### 8.1.3 Details of the Model Solution

In Fig. 8.1, we show plots of the numerical solution of the Hodgkin-Huxley model computed using $\Delta t = 0.001$ ms. In the upper left panel, we have plotted $v$, representing the membrane potential. We see that an action potential is generated, starting with an increase in the value of $v$ (depolarization) followed by a decrease in the value of $v$ (repolarization), like in the simple FitzHugh-Nagumo model studied in Chapter 2. The action potential lasts for a couple of milliseconds.

In the next three panels of Fig. 8.1, we show the value of the unitless $m$, $h$, and $r$ variables, and in the final three panels we show each of the current densities $I_{Na}$, $I_K$, and $I_L$. We see that $I_{Na}$ obtains negative values, while $I_K$ is positive. The current $I_L$ obtains both positive and negative values, and these values are considerably smaller (in absolute value) than the values of $I_{Na}$ and $I_K$ during the action potential. The sign of the current densities can be explained by revisiting the definitions,

$$I_{Na} = g_{Na}m^3 h(v - v_{Na}),$$

$$I_K = g_K r^4 (v - v_K),$$

$$I_L = g_L(v - v_L),$$

(see (8.5)–(8.7)) and recalling that $g_{Na}$, $g_K$, $g_L$, $m$, $h$ and $r$ are all positive. Thus, we see directly from the definition of the current densities that $I_{Na}$ is positive for $v > v_{Na} = 50$ mV and negative for $v < 50$ mV. Similarly, $I_K$ is positive for $v > v_K = -77$ mV and negative for $v < -77$ mV, and $I_L$ is positive for $v > v_L = -54.4$ mV and negative for $v < -54.4$ mV.

Moreover, since $C_m v_t = -(I_{Na} + I_K + I_L)$, see (8.1), and $C_m$ is positive, we deduce that a negative value of the sum ($I_{Na} + I_K + I_L$) is needed for $v_t$ to be positive (i.e., for $v$ to increase) and a positive value of the sum ($I_{Na} + I_K + I_L$) is needed for $v_t$ to be negative (i.e., for $v$ to decrease). We can therefore conclude that in the beginning of the simulation, for $v < -54.4$ mV, both $I_{Na}$ and $I_L$ contribute to the depolarization of $v$, and that after $v$ increases above $-54.4$ mV, $I_{Na}$ is solely responsible for the depolarization. However, as the value of $v$ increases, the value of $(v - v_K)$ and $r$

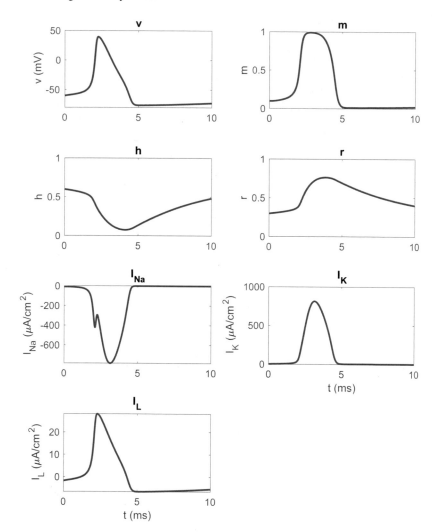

**Fig. 8.1** Numerical solution of the Hodgkin-Huxley model with initial conditions $v(0) = -60$ mV, $m(0) = 0.1, h(0) = 0.6$ and $r(0) = 0.3$. The system of equations is solved using the explicit scheme described in (8.23)–(8.26) with $\Delta t = 0.001$ ms.

increases, which leads to an increased value of $I_K$. When $I_K + I_L$ is more positive than $I_{Na}$ is negative, $v_t$ becomes negative, and $v$ starts to repolarize.

### 8.1.4  Upstroke Velocity and Action Potential Duration

From the discussion in the previous subsection, it seems reasonable to expect that increasing $I_{Na}$, e.g., by increasing the value of the parameter $g_{Na}$, will make the depolarization phase (upstroke) of the action potential more rapid and the repolarization slower (longer action potential duration)[2]. Similarly, we would expect that increasing $I_K$, e.g., by increasing the value of the parameter $g_K$, would make the repolarization of the action potential more rapid, and thus the duration of the action potential shorter. In Fig. 8.2, we show the upstroke and the action potential for a few choices of the parameters $g_{Na}$, $g_K$, and $g_L$. As expected, we observe that increasing $g_{Na}$ increases the upstroke velocity and the action potential duration, while increasing $g_K$ decreases the action potential duration. The adjustments of $g_L$ do not seem to have a significant effect on the computed upstroke or action potential duration.

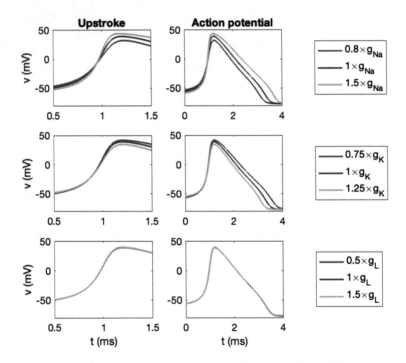

**Fig. 8.2** Numerical solution of the Hodgkin-Huxley model for some adjustments of the parameters $g_{Na}$, $g_K$, and $g_L$. The applied adjustments are given in the legends on the right-hand side of each row, and the remaining parameters are kept at their default values. To make the comparison easier, the timing is adjusted such that the time of the maximal upstroke velocity occurs at the same time for all the parameter choices. The system of equations is solved using the explicit scheme described in (8.23)–(8.26) with $\Delta t = 0.001$ ms.

---

[2] See Section 2.3 (page 17) for definitions of the upstroke velocity and action potential duration.

## 8.2  A Parsimonious Model for the Action Potential of Rabbit Ventricular Cardiomyocytes

In addition to the Hodgkin-Huxley model for neuronal action potentials, we will also consider a similar model for a cardiac action potential. More specifically, we consider a simple, so-called parsimonious, model for the action potential of rabbit ventricular cardiomyocytes from [9]. Note, however, that numerous alternative cardiac action potential models exist (see Section 8.3). The parsimonious rabbit model consists of a system of three ordinary differential equations with three unknowns,

$$C_m \frac{dv}{dt} = -(I_{Na} + I_K + I_{stim}), \tag{8.27}$$

$$\frac{dm}{dt} = \frac{m_\infty - m}{\tau_m}, \tag{8.28}$$

$$\frac{dh}{dt} = \frac{h_\infty - h}{\tau_h}. \tag{8.29}$$

Again, the unknown function $v$ in units of millivolts (mV) represents the membrane potential, and $C_m = 1\ \mu F/cm^2$ is a parameter representing the specific membrane capacitance. Furthermore, $I_{Na}$ and $I_K$ (in $\mu A/cm^2$) represent the current densities through $Na^+$ and $K^+$ channels and are given by

$$I_{Na} = g_{Na} m^3 h(v - v_{Na}), \tag{8.30}$$

$$I_K = g_K e^{-b(v-v_K)}(v - v_K), \tag{8.31}$$

where $g_{Na} = 11\ mS/cm^2$ and $g_K = 0.3\ mS/cm^2$ represent the maximal conductance densities of $Na^+$ and $K^+$ channels, respectively, and $v_{Na} = 65$ mV and $v_K = -83$ mV are the equilibrium potentials of the two channels types. In addition, $m^3 h$ and $e^{-b(v-v_K)}$ represent the open probability of the $Na^+$ and $K^+$ channels, respectively. The parameter $b$ has the value $b = 0.047\ mV^{-1}$. As in the Hodgkin-Huxley model, the unknown functions $m$ and $h$ take values between 0 and 1 and are governed by (8.28)–(8.29)[3] where

$$m_\infty = \frac{1}{1 + e^{(v-E_m)/k_m}}, \qquad \tau_m = 0.12\ ms, \tag{8.32}$$

$$h_\infty = \frac{1}{1 + e^{(v-E_h)/k_h}}, \qquad \tau_h = \frac{2\tau_h^0 e^{\delta_h(v-E_h)/k_h}}{1 + e^{(v-E_h)/k_h}}, \tag{8.33}$$

and

---

[3] Note that the formulation of the equation for $m$ used here ($m_t = (m_\infty - m)/\tau_m$) corresponds to the formulation used in the Hodgkin-Huxley model ($m_t = \alpha_m(1 - m) - \beta_m m$) if we define $\tau_m = \frac{1}{\alpha_m+\beta_m}$ and $m_\infty = \frac{\alpha_m}{\alpha_m+\beta_m}$. The equation for $h$ can also be rewritten analogously.

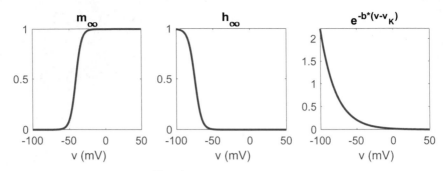

**Fig. 8.3** Values of $m_\infty$, $h_\infty$ and $e^{-b(v-v_K)}$ in the parsimonious ventricular rabbit model (8.27)–(8.35) as functions of $v$.

$$E_m = -41 \text{ mV}, \qquad k_m = -4.0 \text{ mV}, \tag{8.34}$$

$$E_h = -74.9 \text{ mV}, \qquad k_h = 4.4 \text{ mV}, \qquad \tau_h^0 = 6.8 \text{ ms}, \qquad \delta_h = 0.8. \tag{8.35}$$

In addition, $I_{\text{stim}}$ is a stimulus current density given in units of $\mu A/\text{cm}^2$. This stimulus current density is introduced because the initial conditions (see below) yield a system at rest, and the stimulus current is needed to trigger an action potential. The stimulus current is described in more detail in Section 8.2.1 below.

In our computations, we will use the initial conditions

$$v(0) = -83 \text{ mV}, \qquad m(0) = 0, \qquad h(0) = 0.9. \tag{8.36}$$

## 8.2.1 The Stimulus Current Density

In the model (8.27)–(8.29), $I_{\text{stim}}$ is a stimulus current density that is used to initiate an action potential by increasing the membrane potential enough to activate $I_{\text{Na}}$. Specifically, the stimulus current density is given by

$$I_{\text{stim}} = \begin{cases} a_{\text{stim}}, & \text{if } t \geq t_{\text{stim}} \text{ and } t \leq t_{\text{stim}} + d_{\text{stim}}, \\ 0, & \text{otherwise.} \end{cases} \tag{8.37}$$

In other words, the stimulus current density is only nonzero in the period from $t_{\text{stim}}$ to $t_{\text{stim}} + d_{\text{stim}}$. In our computations, we use $a_{\text{stim}} = -25 \ \mu A/\text{cm}^2$, $t_{\text{stim}} = 50$ ms, and $d_{\text{stim}} = 2$ ms, unless otherwise specified. This turns out to be a sufficiently strong $I_{\text{stim}}$ to activate $I_{\text{Na}}$ and thus initiate an action potential.

As mentioned above, without an included stimulus current density, the membrane potential, $v$, is at rest (i.e., does not change with time) at the initial conditions, $v(0) = -83$ mV, $m(0) = 0$, $h(0) = 0.9$. This is because both $I_K$ and $I_{\text{Na}}$ on the right-hand side of (8.27) are equal to or close to zero. The current density $I_K = g_K e^{-b(v-v_K)}(v - v_K)$

is zero because $v = -83$ mV, which is equal to the equilibrium potential of the $K^+$ channels, $v_K$, so $(v - v_K) = 0$. Furthermore, the current density $I_{Na} = g_{Na}m^3h(v - v_{Na})$ is zero because $m$ is zero. From Fig. 8.3, we also see that $m_\infty(v) \approx 0$ for $v = -83$ mV. Therefore, $m_\infty \approx m$, which means that $\frac{dm}{dt} = \frac{m_\infty - m}{\tau_m} \approx 0$ and $m$ is expected to maintain a value close to zero. Since the membrane potential is at rest at $v = -83$ mV, it can be referred to as the *resting potential* of the model.

However, if we apply a negative $I_{stim}$ for some time, this will allow $v$ to increase a bit. If we for example increase $v$ to about $-40$ mV, then $m_\infty(v)$ increases to about 0.5 (see Fig. 8.3), and $\frac{dm}{dt} = \frac{m_\infty - m}{\tau_m}$ becomes positive (if $m \approx 0$), leading to an increased value of $m$. For $m > 0$, we get a nonzero, negative $I_{Na}$, that will last as long as $h > 0$ and $(v - v_{Na}) > 0$ (see (8.30)), and this negative $I_{Na}$ creates the upstroke of the action potential.

## 8.2.2 An Explicit Numerical Scheme

An explicit numerical scheme for the parsimonious ventricular rabbit model can be defined in almost exactly the same manner as in Section 8.1.1 for the Hodgkin-Huxley model by replacing the derivatives in (8.27)–(8.29) by the standard difference (8.18). The computational form for the scheme reads:

$$v_{n+1} = v_n - \frac{\Delta t}{C_m}[I_{Na}(v_n, m_n, h_n) + I_K(v_n) + I_{stim}(t_n)], \qquad (8.38)$$

$$m_{n+1} = m_n + \Delta t \frac{m_\infty(v_n) - m_n}{\tau_m(v_n)}, \qquad (8.39)$$

$$h_{n+1} = h_n + \Delta t \frac{h_\infty(v_n) - h_n}{\tau_h(v_n)}. \qquad (8.40)$$

## 8.2.3 Numerical Computations

In Table 8.2, we report the error, $E_v$, of the numerical solution of $v$ using the numerical scheme (8.38)–(8.40) for some different values of $\Delta t$. The error is computed by comparing the solutions at time $t = 10$ ms to the solutions found using a very fine time step ($\Delta t = 10^{-5}$ ms). As observed for the Hodgkin-Huxley model, we find that the error of the explicit scheme is close to proportional to the time step, $\Delta t$, applied in the numerical scheme. So again we have linear (or first order) convergence.

In Fig. 8.4, we show the numerical solution of the parsimonious ventricular rabbit model solved using $\Delta t = 0.001$ ms. In the upper left plot we show the membrane potential, $v$, and we see that like in Fig. 8.1 for the Hodgkin-Huxley model, an action potential is generated. The action potential starts around the time when $I_{stim}$ is applied (at $t_{stim} = 50$ ms) and seems to last for about 250 ms. This is significantly longer than the about 2-3 ms long neuronal action potential in the Hodgkin-Huxley

**Table 8.2** Error of the numerical solution of the parsimonious ventricular rabbit model for different values of $\Delta t$. The error is defined as $E_v = |v - v_N|$, where $v$ is the numerical solution at $t = 10$ ms for a very fine time step ($\Delta t = 10^{-5}$ ms), and $v_N$ is the numerical solution at $t = 10$ ms for each of the values of $\Delta t$ reported in the first column of the table. In these computations we have used $t_{\text{stim}} = 0$ ms and $d_{\text{stim}} = 2$ ms.

| $\Delta t$ (ms) | $E_v$ (mV) | $E_v/\Delta t$ (mV/ms) |
|---|---|---|
| 0.01 | 0.662 | 66 |
| 0.005 | 0.322 | 64 |
| 0.002 | 0.127 | 63 |
| 0.001 | 0.0627 | 63 |
| 0.0005 | 0.0309 | 62 |

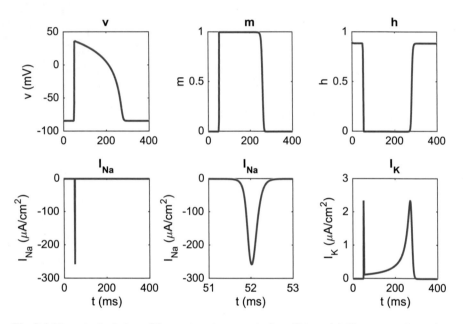

**Fig. 8.4** Numerical solution of the parsimonious ventricular rabbit model. The system of equations is solved using the explicit scheme described in (8.38)–(8.40) with $\Delta t = 0.001$ ms. Note that $I_{\text{Na}}$ is shown in two panels, one with the same time scale as the remaining panels, and one with the time scale zoomed in on the time of the peak current.

model for a neuronal action potential. The next two upper panels show how the value of $m$ and $h$ changes with time, and the plots in the lower panels show the two ion channel current densities, $I_{\text{Na}}$ and $I_{\text{K}}$. The lower left panel shows $I_{\text{Na}}$ in the same time scale as in the remaining panels, and the lower middle panel shows $I_{\text{Na}}$ zoomed in on the points in time when the peak current occurs. The lower right panel shows $I_{\text{K}}$.

### 8.2.4 Upstroke Velocity and Action Potential Duration

In Fig. 8.5, we investigate how the upstroke velocity and action potential duration are affected by adjusting the parameters $g_{Na}$ and $g_K$ in the parsimonious ventricular rabbit model. In the upper panel, we observe that decreasing $g_{Na}$ leads to a slower upstroke and a shorter action potential duration, whereas a lower value of $g_K$ leads to a longer action potential duration. These observations are consistent with what we observed for the Hodgkin-Huxley model in Fig. 8.5. We also observe that the onset of the rapid upstroke happens slightly faster after the stimulus current is applied when $g_K$ is decreased.

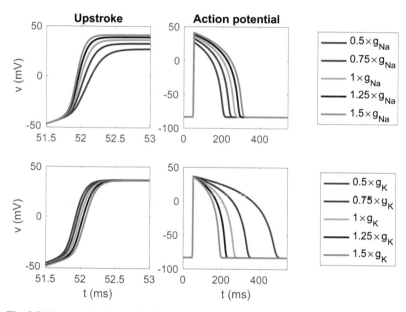

**Fig. 8.5** Numerical solution of $v$ in the parsimonious ventricular rabbit model for some adjustments of the parameters $g_{Na}$ and $g_K$. The applied adjustments are given in the legends on the right-hand side of each row, and the remaining parameters are kept at their default values.. The system of equations are solved using the explicit scheme described in (8.38)–(8.40) with $\Delta t = 0.001$ ms.

## 8.3 Comments and Further Reading

1. Following the publication of the Hodgkin-Huxley model [10] in 1952, many similar models representing the action potential of different cell types were published, building on the same basic principles. One example is the ventricular rabbit model described in Section 8.2. More recent models are often more

complex than the two membrane models considered in this chapter. For instance, a large number of additional membrane currents and dynamic intracellular ion concentrations are often included in the equations. Nevertheless, the resulting system of differential equations can be solved in exactly the same manner as seen for the Hodgkin-Huxley model in Section 8.1.1 and for the parsimonious ventricular rabbit model in Section 8.2.2.

2. A nice overview of the evolution of mathematical membrane models for cardiomyocytes is found in [1]. These models include models for different cell types, like Purkinje fibres (e.g., [4, 14, 19]), ventricular cardiomyocytes (e.g., [7, 16, 20]), and atrial cardiomyocytes (e.g., [3, 8, 15]). They are set up to represent cells from different species, like mice (e.g., [2]), guinea pigs (e.g., [6]), rabbits (e.g., [18]), and humans (e.g., [16]). Membrane models representing stem cell derived cardiomyocytes have also been introduced (e.g., [11, 12, 13, 17]).

3. The membrane models in the form we have considered in this chapter are expressed as a system of ordinary differential equations. The membrane potential, $v$, obtains a single value at each time step and is governed by a number of current densities. There is no spatial variation present in the model. In reality, the currents across the membrane happen through ion channels at different locations of the membrane. The current densities in the models in the form considered in this chapter could therefore be interpreted as the average current densities over a given area of membrane, for example, over the membrane of one cell. In this case, $v$ could be interpreted as the averaged membrane potential of the cell. In the next chapters, we will combine the membrane models considered in this chapter with spatial models of electrophysiology.

4. The formulation of the Hodgkin-Huxley model given in Section 8.1 is adjusted for the membrane potential to have a resting state at $v = -65$ mV, and is taken from [5].

5. Compared to the original publication of the parsimonious ventricular rabbit model [9], we have for simplicity used values of $\tau_h^0$ and $\delta_h$ rounded off to one decimal point.

6. A cardiomyocyte action potential typically lasts several hundred milliseconds. In the parsimonious model considered above, it typically lasts between 200 ms and 500 ms (see Fig. 8.5). In Fig. 8.4, we note that the gradients of the solution are very sharp, and this indicates that we need to apply small time steps to pick up the main features of the solutions. From Table 8.2, we note that the error is about $E_v \approx 63 \times \Delta t$ mV/ms. Thus, if we accept an error of about 1 mV, we need the time step to be less that 1/63 ms. If the action potential lasts for 500 ms, we need to perform at least 31,500 time steps to achieve sufficient accuracy.

7. The action potential of a neuron lasts only a few milliseconds, while that of a cardiomyocyte lasts several hundred milliseconds. The difference in duration is due to a difference in the density of $Na^+$ and $K^+$ channels in the membrane of the two cell types. In the models considered in this chapter, the neuronal model has a 10 times higher density of $Na^+$ channels and around a 100 times higher density of $K^+$ channels than the cardiac model. The high density of channels leads to a brief and clear signal transmission in the neuronal model. On the other hand, the

longer action potential of the cardiomyocyte is crucial in inducing an increase in intracellular $Ca^{2+}$ concentration (not part of the model above), resulting in the mechanical contraction of the myocyte and effective pumping of blood. Thus, the longer action potential is essential for optimal cardiac function.

# References

[1] Amuzescu B, Airini R, Epureanu FB, Mann SA, Knott T, Radu BM (2021) Evolution of mathematical models of cardiomyocyte electrophysiology. Mathematical Biosciences 334:108567

[2] Bondarenko VE, Szigeti GP, Bett GC, Kim SJ, Rasmusson RL (2004) Computer model of action potential of mouse ventricular myocytes. American Journal of Physiology-Heart and Circulatory Physiology 287(3):H1378–H1403

[3] Courtemanche M, Ramirez RJ, Nattel S (1998) Ionic mechanisms underlying human atrial action potential properties: insights from a mathematical model. American Journal of Physiology-Heart and Circulatory Physiology 275(1):H301–H321

[4] Di Francesco D, Noble D (1985) A model of cardiac electrical activity incorporating ionic pumps and concentration changes. Philosophical Transactions of the Royal Society of London B, Biological Sciences 307(1133):353–398

[5] Ermentrout GB, Terman DH (2010) Mathematical foundations of neuroscience, vol 35. Springer

[6] Faber GM, Rudy Y (2000) Action potential and contractility changes in $[Na^+]$ i overloaded cardiac myocytes: a simulation study. Biophysical Journal 78(5):2392–2404

[7] Grandi E, Pasqualini FS, Bers DM (2010) A novel computational model of the human ventricular action potential and Ca transient. Journal of Molecular and Cellular Cardiology 48(1):112–121

[8] Grandi E, Pandit SV, Voigt N, Workman AJ, Dobrev D, Jalife J, Bers DM (2011) Human atrial action potential and $Ca^{2+}$ model: sinus rhythm and chronic atrial fibrillation. Circulation Research 109(9):1055–1066

[9] Gray RA, Pathmanathan P (2016) A parsimonious model of the rabbit action potential elucidates the minimal physiological requirements for alternans and spiral wave breakup. PLoS Computational Biology 12(10):e1005087

[10] Hodgkin AL, Huxley AF (1952) A quantitative description of membrane current and its application to conduction and excitation in nerve. The Journal of Physiology 117(4):500–544

[11] Jæger KH, Charwat V, Charrez B, Finsberg H, Maleckar MM, Wall S, Healy KE, Tveito A (2020) Improved computational identification of drug response using optical measurements of human stem cell derived cardiomyocytes in microphysiological systems. Frontiers in Pharmacology 10:1648

[12] Jæger KH, Wall S, Tveito A (2021) Computational prediction of drug response in short QT syndrome type 1 based on measurements of compound effect in stem cell-derived cardiomyocytes. PLoS Computational Biology 17(2):e1008089

[13] Kernik DC, Morotti S, Wu H, Garg P, Duff HJ, Kurokawa J, Jalife J, Wu JC, Grandi E, Clancy CE (2019) A computational model of induced pluripotent stem-cell derived cardiomyocytes incorporating experimental variability from multiple data sources. The Journal of Physiology 597(17):4533–4564

[14] Noble D (1962) A modification of the Hodgkin–Huxley equations applicable to purkinje fibre action and pacemaker potentials. The Journal of Physiology 160(2):317–352

[15] Nygren A, Fiset C, Firek L, Clark JW, Lindblad DS, Clark RB, Giles WR (1998) Mathematical model of an adult human atrial cell: the role of K+ currents in repolarization. Circulation Research 82(1):63–81

[16] O'Hara T, Virág L, Varró A, Rudy Y (2011) Simulation of the undiseased human cardiac ventricular action potential: model formulation and experimental validation. PLoS Computational Biology 7(5):e1002061

[17] Paci M, Hyttinen J, Aalto-Setälä K, Severi S (2013) Computational models of ventricular-and atrial-like human induced pluripotent stem cell derived cardiomyocytes. Annals of Biomedical Engineering 41(11):2334–2348

[18] Shannon TR, Wang F, Puglisi J, Weber C, Bers DM (2004) A mathematical treatment of integrated Ca dynamics within the ventricular myocyte. Biophysical Journal 87(5):3351–3371

[19] Stewart P, Aslanidi OV, Noble D, Noble PJ, Boyett MR, Zhang H (2009) Mathematical models of the electrical action potential of purkinje fibre cells. Philosophical Transactions of the Royal Society A: Mathematical, Physical and Engineering Sciences 367(1896):2225–2255

[20] ten Tusscher KH, Panfilov AV (2006) Alternans and spiral breakup in a human ventricular tissue model. American Journal of Physiology-Heart and Circulatory Physiology 291(3):H1088–H1100

# Chapter 9
# The Cable Equation

In Chapter 6, we studied a simple version of the cable equation, where a diffusion term was added to the FitzHugh-Nagumo equations. In this chapter, we will revisit the cable equation and go through a simple derivation of the model. In addition, we will consider the numerical solution of the cable equation for a neuronal axon with membrane dynamics modeled by the Hodgkin-Huxley model.

## 9.1 Derivation of the Cable Equation

In order to get a sense of the origin of the terms in the cable equation, we will here consider a simple derivation of the model. Similar derivations are found in, e.g., [1, 2, 4, 7, 9]. The derivation is based on dividing a cell (e.g., an axon) into a number of compartments in the $x$-direction, as illustrated in Fig. 9.1.

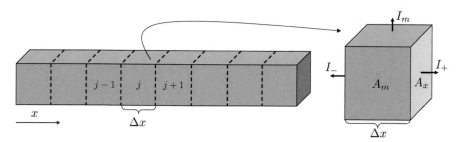

**Fig. 9.1** Left: Illustration of a cell separated into a number of compartments of length $\Delta x$. Right: Illustration of one of the compartments, $j$. The cross-sectional area in the $x$-direction is denoted by $A_x$, and the total membrane area is denoted by $A_m$. The current across the membrane is denoted by $I_m$, the current from compartment $j$ to compartment $j + 1$ is denoted by $I_+$ and the current from compartment $j$ to compartment $j - 1$ is denoted by $I_-$.

K. Horgmo Jæger, A. Tveito, *Differential Equations for Studies in Computational Electrophysiology*,
Simula SpringerBriefs on Computing 14, https://doi.org/10.1007/978-3-031-30852-9_9

The left panel of Fig. 9.1 illustrates a cell divided into a number of compartments of length $\Delta x$, and the right panel illustrates an arbitrary inner[1] compartment number $j$. We will derive the cable equation by considering each of the electric currents flowing out of this compartment.

## Currents Between Compartments

We assume that the currents that flow between compartments are governed by Ohm's law in the sense that the current from compartment $j$ to compartment $j + 1$ is given by

$$I_+ = \frac{u_{i,j} - u_{i,j+1}}{R}, \tag{9.1}$$

where $u_{i,j}$ is the electrical potential the center of compartment $j$ and $u_{i,j+1}$ is the electrical potential in the center of compartment $j + 1$, both in units of millivolts (mV). The subscript $i$ is used to specify that we are considering the intracellular potential of the cell. Furthermore, $R$ is the intracellular resistance between the two compartment centers in units of kilo-Ohm's (k$\Omega$). This resistance can be expressed as (see, e.g., [5])

$$R = \frac{\Delta x}{\sigma_i A_x}, \tag{9.2}$$

where $\Delta x$ is the distance between the centers of compartments $j$ and $j + 1$ (in cm), $\sigma_i$ is the intracellular conductivity (in mS/cm), and $A_x$ (in cm$^2$) is the cross sectional area of the cell (see Fig. 9.1). Inserting (9.2) into (9.1), we get

$$I_+ = \sigma_i A_x \frac{u_{i,j} - u_{i,j+1}}{\Delta x}, \tag{9.3}$$

and, following the same steps, we get that the current from compartment $j$ to compartment $j - 1$ is given by

$$I_- = \sigma_i A_x \frac{u_{i,j} - u_{i,j-1}}{\Delta x}. \tag{9.4}$$

The unit of $I_+$ and $I_-$ is micro-Amperes ($\mu$A).

## Membrane Currents

We assume that the membrane acts as a capacitor that can store charge and that this property can be modeled like in the membrane models in Chapter 8. More specifically, we assume that the total current across the membrane of compartment

---

[1] Here, an arbitrary inner compartments refers to any of the compartments except for the rightmost or leftmost compartments.

$j$ is given by[2]

$$I_m = A_m \left( C_m \frac{\partial v_j}{\partial t} + I_{\text{ion}} \right). \tag{9.5}$$

Here, $A_m$ is the membrane area of compartment $j$ (in cm$^2$), given by

$$A_m = \eta_m \Delta x, \tag{9.6}$$

where $\eta_m$ is the circumference of the compartment (in cm) and $\Delta x$ is the length of the compartment (in cm). Furthermore, $C_m$ is the specific membrane capacitance (given in units of $\mu$F/cm$^2$), and $v_j$ is the membrane potential (in mV) defined as the potential difference

$$v = u_i - u_e, \tag{9.7}$$

where $u_i$ is the intracellular potential in the compartment and $u_e$ is the extracellular potential outside of the compartment. The term $C_m \frac{\partial v_j}{\partial t}$ is called the capacitive current density and represents the current density to and from the collection of charges stored by the membrane capacitor. Moreover, $I_{\text{ion}}$ is the current density (in $\mu$A/cm$^2$) through ion channels in the cell membrane. For example, if we assume that the membrane dynamics are modeled by the Hodgkin-Huxley model (see Section 8.1), this current density is given by

$$I_{\text{ion}} = I_{\text{Na}} + I_K + I_L. \tag{9.8}$$

**Sum of Currents**

In order to derive the cable equation, we assume that Kirchhoff's current law applies in each compartment. In other words, we assume that the sum of all of the currents out of the compartment is zero. This gives

$$I_+ + I_- + I_m = 0, \tag{9.9}$$

and inserting (9.3)–(9.6), this yields

$$\sigma_i A_x \frac{u_{i,j} - u_{i,j+1}}{\Delta x} + \sigma_i A_x \frac{u_{i,j} - u_{i,j-1}}{\Delta x} + \eta_m \Delta x \left( C_m \frac{\partial v_j}{\partial t} + I_{\text{ion}} \right) = 0, \tag{9.10}$$

which can be rearranged to

$$C_m \frac{\partial v_j}{\partial t} = \frac{\sigma_i A_x}{\eta_m} \frac{u_{i,j-1} - 2u_{i,j} + u_{i,j+1}}{\Delta x^2} - I_{\text{ion}}. \tag{9.11}$$

---

[2] In (9.5) we recognize the terms $C_m v_t$ and $I_{\text{ion}}$ (see (9.8)) from the membrane models considered in Chapter 8. In those models, we ignored all spatial variation and assumed that $I_m$ was the only current into or out of the cell. In order for the sum of the currents to be zero, we therefore ended up with models on the form $C_m v_t = -I_{\text{ion}}$.

Now, we insert an assumption that the extracellular potential is zero everywhere such that $v = u_i$ (see (9.7))[3]. In that case, the model reads

$$C_m \frac{\partial v_j}{\partial t} = \delta \frac{v_{j-1} - 2v_j + v_{j+1}}{\Delta x^2} - I_{\text{ion}}, \tag{9.12}$$

where

$$\delta = \frac{\sigma_i A_x}{\eta_m}. \tag{9.13}$$

From Chapter 3 (see page 23), we recall that

$$v_{xx}(t, x) \approx \frac{v(t, x - \Delta x) - 2v(t, x) + v(t, x + \Delta x)}{\Delta x^2} \tag{9.14}$$

for a sufficiently small $\Delta x$. We therefore assume that $\Delta x$ in (9.12) is very small, and obtain the cable equation

$$C_m \frac{\partial v}{\partial t} = \delta \frac{\partial^2 v}{\partial x^2} - I_{\text{ion}}. \tag{9.15}$$

Since the considered compartment $j$ was chosen as an arbitrary inner compartment, we conclude that the equation (9.15) applies everywhere along the cell, except at the left and right boundaries. The boundary conditions are considered in the next subsection.

### 9.1.1  Boundary Conditions

Intuitively, for the leftmost compartment of the cell (see Fig. 9.1), the current from this compartment to the *non-existing* compartment to the left, $I_-$, should be assumed to be zero. In other words,

$$I_- = \sigma_i A_x \frac{u_{i,j} - u_{i,j-1}}{\Delta x} = 0. \tag{9.16}$$

From previous chapters (see, e.g., (1.6) in Chapter 1), we know that

$$\frac{\partial u_i}{\partial x}(t, x) \approx \frac{u_i(t, x) - u_i(t, x - \Delta x)}{\Delta x} \tag{9.17}$$

for a sufficiently small $\Delta x$. Assuming that $\Delta x$ is very small, (9.16) can therefore be translated to the boundary condition

$$\sigma_i A_x \frac{\partial u_i}{\partial x}(t, 0) = 0, \tag{9.18}$$

---

[3] Note that the assumption that the extracellular potential is zero could be replaced by alternative assumptions (see the Section 9.4 for more details).

where we have assumed that the leftmost boundary of the cell is located at $x = 0$. Dividing by $\sigma_i A_x$ on both sides of the equation and using the assumption that the extracellular potential is zero, which gives $u_i = v$, we get the boundary condition

$$\frac{\partial v}{\partial x}(t,0) = 0. \tag{9.19}$$

Inserting this into the discrete version of the cable equation for the left boundary, we get

$$C_m \frac{\partial v_1}{\partial t} = \delta \frac{-v_1 + v_2}{\Delta x^2} - I_{\text{ion}}. \tag{9.20}$$

A similar argument for the right boundary of the cell, at $x = L$, gives the boundary condition

$$\frac{\partial v}{\partial x}(t,L) = 0 \tag{9.21}$$

and the discrete equation

$$C_m \frac{\partial v_M}{\partial t} = \delta \frac{-v_M + v_{M-1}}{\Delta x^2} - I_{\text{ion}}. \tag{9.22}$$

## 9.1.2 Geometry

The value of $\delta$ (see (9.13)) in the cable equation depends on the considered geometry. If we consider a rectangular cuboid, as illustrated in Fig. 9.1, with width $w$ in the $y$- and $z$-directions, we have $A_x = w^2$ and $\eta_m = 4w$, which gives

$$\delta = \frac{\sigma_i A_x}{\eta_m} = \frac{w\sigma_i}{4}. \tag{9.23}$$

Similarly, if we consider a cylinder with radius $r$, we have $A_x = \pi r^2$ and $\eta_m = 2\pi r$, which gives

$$\delta = \frac{\sigma_i A_x}{\eta_m} = \frac{r\sigma_i}{2}. \tag{9.24}$$

## 9.1.3 Additional State Variables

As mentioned in the derivation above, the term $I_{\text{ion}}$ represents the current density through ion channels in the cell membrane. This current density could, for example, be modeled by the Hodgkin-Huxley model:

$$I_{\text{ion}} = I_{\text{Na}} + I_{\text{K}} + I_{\text{L}} \tag{9.25}$$

(see Section 8.1). Here, the formulation of $I_{Na}$ and $I_K$ involve the additional state variables $m$, $h$ and $r$, which are governed by (8.2)–(8.4). In order to define the current density, $I_{ion}$, we therefore need to include these equations in the model, where $m$, $h$ and $r$ are functions of both $t$ and $x$. We thus get a system of equations of the form

$$C_m \frac{\partial v}{\partial t} = \delta \frac{\partial^2 v}{\partial x^2} - I_{ion}, \tag{9.26}$$

$$\frac{\partial m}{\partial t} = \alpha_m(1 - m) - \beta_m m, \tag{9.27}$$

$$\frac{\partial h}{\partial t} = \alpha_h(1 - h) - \beta_h h, \tag{9.28}$$

$$\frac{\partial r}{\partial t} = \alpha_r(1 - r) - \beta_r r. \tag{9.29}$$

## 9.2 Numerical Schemes

We will consider two alternative numerical schemes for the cable equation. First, a straightforward explicit scheme and then an operator splitting scheme, treating the diffusion part of the system implicitly. In both schemes, we seek numerical approximations to the solution in the $M$ spatial points $x_j = (j - 1) \times \Delta x$, for $j = 1, ..., M$, at the time points $t_n = n \times \Delta t$ for $n = 1, ..., N$. Here, $\Delta x = \frac{L}{M-1}$ and $\Delta t = \frac{T}{N}$, where $L$ is the length of the domain and $T$ is the total simulation time.

### 9.2.1 An Explicit Numerical Scheme for the Cable Equation

We consider a numerical scheme for the cable equation that is based on this discrete spatial version of the equation considered in the derivation of the model, i.e., (9.12). To define a numerical scheme, we have to replace the remaining derivative $\frac{\partial v}{\partial t}$ by a difference, and we choose the usual difference (see, e.g., (1.5) on page 4). In addition, similar replacements of derivatives by differences are inserted into (9.27)–(9.29), and we get the explicit scheme

$$C_m \frac{v_j^{n+1} - v_j^n}{\Delta t} = \delta \frac{v_{j-1}^n - 2v_j^n + v_{j+1}^n}{\Delta x^2} - I_{ion}(v_j^n, m_j^n, h_j^n, r_j^n), \tag{9.30}$$

$$\frac{m_j^{n+1} - m_j^n}{\Delta t} = \alpha_m(v_j^n)(1 - m_j^n) - \beta_m(v_j^n)m_j^n, \tag{9.31}$$

$$\frac{h_j^{n+1} - h_j^n}{\Delta t} = \alpha_h(v_j^n)(1 - h_j^n) - \beta_h(v_j^n)h_j^n, \tag{9.32}$$

$$\frac{r_j^{n+1} - r_j^n}{\Delta t} = \alpha_r(v_j^n)(1 - r_j^n) - \beta_r(v_j^n)r_j^n, \tag{9.33}$$

for $j = 2, ..., M - 1$. For $j = 1$ and $j = M$, we replace the right-hand side of (9.30) by the right-hand sides of (9.20) and (9.22), respectively. This scheme can be rewritten to computational form

$$v^{n+1} = \left(I + \frac{\Delta t}{C_m} A\right) v^n - \frac{\Delta t}{C_m} I_{\text{ion}}(v^n, m^n, h^n, r^n), \tag{9.34}$$

$$m^{n+1} = m^n + \Delta t[\alpha_m(v^n)(1 - m^n) - \beta_m(v^n)m^n], \tag{9.35}$$

$$h^{n+1} = h^n + \Delta t[\alpha_h(v^n)(1 - h^n) - \beta_h(v^n)h^n], \tag{9.36}$$

$$r^{n+1} = r^n + \Delta t[\alpha_r(v^n)(1 - r^n) - \beta_r(v^n)r^n], \tag{9.37}$$

where

$$v^n = \begin{pmatrix} v_1^n \\ v_2^n \\ \vdots \\ v_M^n \end{pmatrix}, \quad m^n = \begin{pmatrix} m_1^n \\ m_2^n \\ \vdots \\ m_M^n \end{pmatrix}, \quad h^n = \begin{pmatrix} h_1^n \\ h_2^n \\ \vdots \\ h_M^n \end{pmatrix}, \quad r^n = \begin{pmatrix} r_1^n \\ r_2^n \\ \vdots \\ r_M^n \end{pmatrix}, \tag{9.38}$$

the matrix $A$ is defined by

$$A = \frac{\delta}{\Delta x^2} \begin{pmatrix} -1 & 1 & 0 & \cdots & & & 0 \\ 1 & -2 & 1 & 0 & \cdots & & 0 \\ 0 & 1 & -2 & 1 & 0 & \cdots & 0 \\ \vdots & & \ddots & \ddots & \ddots & & \vdots \\ 0 & \cdots & & & 0 & 1 & -1 \end{pmatrix}, \tag{9.39}$$

and $I$ is the $M \times M$ identity matrix.

## 9.2.2 An Operator Splitting Scheme for the Cable Equation

A potential disadvantage of the explicit scheme defined above is instability issues, like observed in Chapter 3 (see Section 3.4.3). In an attempt to avoid these issues, we therefore also define an operator splitting scheme for the cable equation, with an implicit treatment of the diffusion part of the system. In other words, we divide the system into two parts for every time step by first solving

$$C_m \frac{\partial v}{\partial t} = \delta \frac{\partial^2 v}{\partial x^2}, \quad \frac{\partial m}{\partial t} = 0, \quad \frac{\partial h}{\partial t} = 0, \quad \frac{\partial r}{\partial t} = 0, \tag{9.40}$$

using an implicit scheme and then solving

$$C_m \frac{\partial v}{\partial t} = -I_{\text{ion}}, \tag{9.41}$$

$$\frac{\partial m}{\partial t} = \alpha_m (1 - m) - \beta_m m, \tag{9.42}$$

$$\frac{\partial h}{\partial t} = \alpha_h (1 - h) - \beta_h h, \tag{9.43}$$

$$\frac{\partial r}{\partial t} = \alpha_r (1 - r) - \beta_r r, \tag{9.44}$$

using an explicit scheme. In computational form, using the vectors $v_n$, $m_n$, $h_n$, $r_n$ as defined in (9.38) and the matrix $A$ as defined in (9.39), the scheme for the first step reads

$$\left(I - \frac{\Delta t}{C_m} A\right) v^{n+1/2} = v^n, \quad m^{n+1/2} = m^n, \quad h^{n+1/2} = h^n, \quad r^{n+1/2} = r^n, \tag{9.45}$$

and the scheme for the second step reads

$$v^{n+1} = v^{n+1/2} - \frac{\Delta t}{C_m} I_{\text{ion}}(v^{n+1/2}, m^{n+1/2}, h^{n+1/2}, r^{n+1/2}), \tag{9.46}$$

$$m^{n+1} = m^{n+1/2} + \Delta t [\alpha_m(v^{n+1/2})(1 - m^{n+1/2}) - \beta_m(v^{n+1/2})m^{n+1/2}], \tag{9.47}$$

$$h^{n+1} = h^{n+1/2} + \Delta t [\alpha_h(v^{n+1/2})(1 - h^{n+1/2}) - \beta_h(v^{n+1/2})h^{n+1/2}], \tag{9.48}$$

$$r^{n+1} = r^{n+1/2} + \Delta t [\alpha_r(v^{n+1/2})(1 - r^{n+1/2}) - \beta_r(v^{n+1/2})r^{n+1/2}]. \tag{9.49}$$

## 9.3 Numerical Computations

We apply the two numerical schemes to a simple test case for the cable equation with membrane dynamics modeled by the Hodgkin-Huxley model. We consider a cell shaped as a rectangular cuboid with length $L = 0.5$ cm and width $w = 0.001$ cm, and we assume that $\sigma_i = 4$ mS/cm. This gives $\delta = 0.001$ mS (see (9.13)). Furthermore, the initial conditions are given by

$$v(0, x) = \begin{cases} -50 \text{ mV, for } x <= 0.05 \text{ cm,} \\ -65 \text{ mV, for } x > 0.05 \text{ cm,} \end{cases} \tag{9.50}$$

$$m(0, x) = 0.1, \quad h(0, x) = 0.6, \quad r(0, x) = 0.3. \tag{9.51}$$

### 9.3.1 Stability

In Fig. 9.2, we show the numerical solution of the problem found using the explicit scheme with $\Delta x = 0.001$ cm and $\Delta t = 0.001$ ms at some different points in time. We

**Fig. 9.2** Numerical solution of $v$ in the cable equation with membrane dynamics modeled by the Hodgkin-Huxley model at five different points in time. The numerical solution is found using the explicit numerical scheme (9.34)–(9.37), with $\Delta x = 0.001$ cm and $\Delta t = 0.001$ ms. We have zoomed in on the solution at the leftmost part of the domain, and observe that we get unreasonable oscillations in the area close to the discontinuity of the initial conditions ($x = 0.05$ cm).

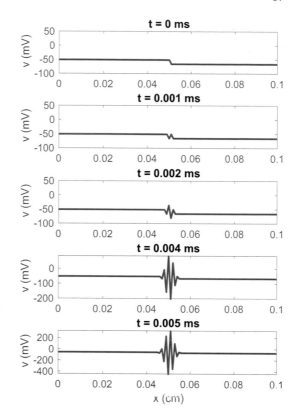

observe that already at the first time steps, we get unreasonable oscillations close to the discontinuity in the initial conditions (at $x = 0.05$ cm, see (9.50)).

In Fig. 9.3, we have solved the system of equations using the explicit scheme with a smaller value of $\Delta t$ ($\Delta t = 0.0002$ ms). In this case, we avoid the oscillations, and we seem to get a reasonable traveling wave solution. However, because of the small time step, the numerical computations are quite slow. For this reason, we also try to solve the system using the operator splitting scheme described in Section 9.2.2. Using this scheme, we are able to use a time step of $\Delta t = 0.02$ ms, and still get a solution that is very similar to the one that we got using the explicit scheme with a fine time step (compare the solid blue and dotted orange lines in Fig. 9.3). Even though we have to solve a linear system of equations in order to find the solution in the first step of the operator splitting scheme, the algorithm is able to find the solution much faster than the explicit scheme because of the large difference in the required time step. More specifically, for the laptop computer that we used to perform the computations, the operator splitting scheme was about 25 times faster than the explicit scheme.

**Fig. 9.3** Numerical solution of $v$ in the cable equation with membrane dynamics modeled by the Hodgkin-Huxley model at five different points in time. The solution drawn with a solid line is found using the explicit numerical scheme (9.34)–(9.37), with $\Delta x = 0.001$ cm and $\Delta t = 0.0002$ ms, and the solution drawn with a dotted line is found using the operator splitting scheme described in Section 9.2.2, with $\Delta x = 0.001$ cm and $\Delta t = 0.02$ ms.

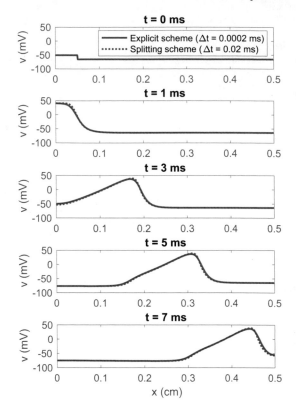

## 9.3.2 Conduction Velocity

In Chapter 6, we considered a unitless version of the cable equation with membrane dynamics modeled by the FitzHugh-Nagumo model and observed how the conduction velocity, CV (i.e., the velocity with which the traveling wave moved though the domain) depended on two of the model parameters. In Fig. 9.4, we show the results of a similar experiment for the cable equation with membrane dynamics modeled by the Hodgkin-Huxley model. We vary the value of the three parameters $g_{Na}$, $g_K$, and $g_L$ representing the maximal conductance density of the three membrane currents of the Hodgkin-Huxley model (see (8.5)–(8.7)). In addition, we vary the value of the cell width, $w$, and the intracellular conductivity, $\sigma_i$, used to define $\delta$ in the cable equation (see (9.23)).

We define the conduction velocity as

$$CV = \frac{x_2 - x_1}{t_2 - t_1}, \tag{9.52}$$

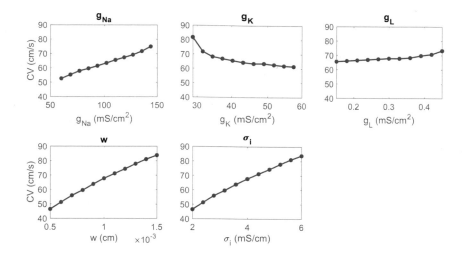

**Fig. 9.4** Conduction velocity (CV) computed from the numerical solution of the cable equation as described in (9.52) for a few adjustments of parameters. The parameter values not explicitly stated on the $x$-axis are set to their default values. The numerical solution is found using the operator splitting scheme described in Section 9.2.2, with $\Delta x = 0.001$ cm and $\Delta t = 0.02$ ms.

where $x_1 = 0.2$ cm and $x_2 = 0.4$. Furthermore, $t_1$ and $t_2$ are the points in time when the value of $v$ first increases to a value $v > 0$ mV, in the spatial points $x_1$ and $x_2$, respectively.

In Fig. 9.4, we observe that increasing the value of $g_{Na}$, $\sigma_i$ or the cell width, $w$, significantly increases the conduction velocity. Increasing $g_L$ also increases the conduction velocity somewhat, whereas increasing $g_K$ decreases the conduction velocity.

## 9.4  Comments and Further Reading

1. In the derivation of the cable equation given in this chapter, we assume that the extracellular potential is zero. However, if we instead assume that the extracellular potential is another constant, $C$, different from zero, we end up with the exact same formulation of the cable equation. In that case, we would have $u_i^j = v^j + C$ (see (9.7)), and inserting this into (9.11), we get

$$C_m \frac{\partial v^j}{\partial t} = \delta \frac{v^{j-1} + C - 2(v^j + C) + v^j + C}{\Delta x^2} - I_{ion}$$

$$= \delta \frac{v^{j-1} - 2v^j + v^j}{\Delta x^2} - I_{ion}. \qquad (9.53)$$

which is exactly the same as (9.12).

2. The cable equation can also be derived by assuming that the value of the extracellular potential can vary in the $x$-direction, but not in the $y$- and $z$-directions. This leads to a different formulation of the constant $\delta$ in the cable equation. See, e.g., [4] for more details on this version of the cable equation.

3. In Section 9.1.1, we derived boundary conditions for the cable equation by considering the leftmost and rightmost compartments of the cell. However, since it might be reasonable to assume that the left and right sides of the cell are covered by membrane (see Fig. 9.1), we might also wish to assume that the amount of membrane area is larger for these compartments. More specifically, for the leftmost and rightmost compartments, we get

$$A_m = A_x + A_m = \Delta x \left( \frac{A_x}{\Delta x} + \eta_m \right). \tag{9.54}$$

Inserting this into the discrete version of the cable equation, we get the adjusted

$$\delta_b = \frac{\sigma_i A_x}{\frac{A_x}{\Delta x} + \eta_m} \tag{9.55}$$

for the left and right boundaries. In the code associated with these notes, we have included an example simulation where we compare the solution of the system using this adjusted $\delta_b$ at the boundary to the case where the default $\delta$ is used everywhere. By running that code, we see that the solutions for these two cases are indistinguishable.

4. In this chapter, we considered the cable equation for modeling the spread of an electrical signal along a neuronal axon. However, the cable equation (9.15) can also be used to study the spread of an electrical signal along a cardiac fibre composed of a row of cardiomyocytes connected to each other by gap junctions. In this case, the increased resistance for the currents across the gap junctions has to be taken into account, either in an averaged manner (see, e.g., [6]) or by having a $\delta$ that depends on $x$ in a way such that the discretely located cell-to-cell connections are represented (see, e.g., [3]).

5. In (9.10) we derived a compartmental version of the cable model equations. Then, we passed to the limit in $\Delta x$ and obtained the differential equation (9.14). Finally, we replaced the derivatives by differences, and obtained the scheme (9.30) which is, more or less, the compartmental model we started by deriving. So, was the differential equation just a detour? Do we need it? No, we don't really need it, but it is customary to phrase models in terms of partial differential equations. The main reason for this may be tradition, but there are practical reasons as well. When the problem is formulated as a partial differential equation, we have a large collection of methods that can be applied; finite difference methods, finite volume methods, finite element methods, boundary element methods, and many more. Also, the problem can be expressed in a very compact manner using differential equations whereas the compartmental form (or finite difference form for that matter) is clunky and much harder to read for complex systems involving

many equations. Finally, differential equations can be analyzed mathematically using tools that are harder to apply when the equations are discretized. But the major disadvantage of the formulation as a differential equation is that it is not straightforward to solve using computers – the form and popularity is inherited from a time where math was done using paper and pencils. An attempt to introduce partial differential equations using both continuous and discrete approaches can be found in [8].

# References

[1] Einevoll GT (2006) Mathematical modeling of neural activity. In: Dynamics of Complex Interconnected Systems: Networks and Bioprocesses, Springer, pp 127–145

[2] Ermentrout GB, Terman DH (2010) Mathematical foundations of neuroscience, vol 35. Springer

[3] Henriquez CS (2014) A brief history of tissue models for cardiac electrophysiology. IEEE Transactions on Biomedical Engineering 61(5):1457–1465

[4] Keener J, Sneyd J (2009) Mathematical physiology 1: Cellular physiology, vol 2. Springer New York, NY, USA

[5] Plonsey R, Barr RC (2007) Bioelectricity: a quantitative approach. Springer

[6] Shaw RM, Rudy Y (1997) Ionic mechanisms of propagation in cardiac tissue: roles of the sodium and L-type calcium currents during reduced excitability and decreased gap junction coupling. Circulation Research 81(5):727–741

[7] Sterratt D, Graham B, Gillies A, Willshaw D (2011) Principles of computational modelling in neuroscience. Cambridge University Press

[8] Tveito A, Winther R (2009) Introduction to partial differential equations; a computational approach, 2nd edn. Springer

[9] Tveito A, Jæger KH, Lines GT, Paszkowski Ł, Sundnes J, Edwards AG, Māki-Marttunen T, Halnes G, Einevoll GT (2017) An evaluation of the accuracy of classical models for computing the membrane potential and extracellular potential for neurons. Frontiers in Computational Neuroscience 11:27

# Chapter 10
# Spatial Models of Cardiac Electrophysiology

In Chapter 8, we introduced mathematical models of the action potential across a cell membrane. Understanding the properties of the cell membrane is absolutely essential in order to understand the electrophysiology of excitable cells. However, some essential properties can only be studied in spatially resolved models; i.e., in models representing spatial variation across a single cell or a collection of cells.

Every heartbeat is based on an electrochemical wave traversing the whole cardiac muscle. Strong perturbations to these waves are referred to as arrhythmias. These disturbances can seriously disrupt the contraction of the heart muscle and are therefore very dangerous and potentially lethal. Cardiac fibrillation refers to a state of the heart where the contraction is completely unsynchronized, leading to severely reduced pumping functions. In mathematical modelling, fibrillation can only be studied in spatially resolved models, and hence the membrane models introduced above are inadequate. But these models are excellent building blocks in models representing collections of cells.

A first step towards spatially resolved modeling was presented in Chapter 9, where we discussed the cable equation. This model is often used to study a strand of cardiac cells, but since the model is inherently one-dimensional, it has limited relevance for complex spatial phenomena like cardiac fibrillation. Here, we will present the celebrated bidomain model. It is considered to represent the gold standard of mathematical models of cardiac electrophysiology and dates back 50 years. From the bidomain model it is easy to derive the somewhat simpler monodomain model. This model is easier to deal with in terms of numerical solution and is therefore frequently used as a replacement of the more correct bidomain model.

Here we will just present the models and show that the techniques we have derived above can be used to obtain numerical solutions of both the monodomain model and the bidomain model. In the notes below we will point to literature where the models are derived, and we will point to better numerical methods and to applications of the models. Again, we will just consider the finite difference method in order to keep things as simple as possible (but not simpler) and therefore we will use two-dimensional squares as computational domains. However, excellent open-source software tools are available for numerical simulations based on the

© The Author(s) 2023
K. Horgmo Jæger, A. Tveito, *Differential Equations for Studies in Computational Electrophysiology*,
Simula SpringerBriefs on Computing 14, https://doi.org/10.1007/978-3-031-30852-9_10

bidomain and monodomain models using the finite element method. Using the finite element method, more realistic geometries can be applied.

## 10.1 First, the Diffusion Equation, Again

For purely technical and notational reasons, we will first briefly consider the two-dimensional diffusion equation. This is helpful in order to understand the mesh, the matrices and the vectors that we use to solve the bidomain and monodomain equations below.

Consider the following initial and boundary value problem (unitless),

$$\frac{\partial u(t,x,y)}{\partial t} = \sigma \frac{\partial^2 u(t,x,y)}{\partial x^2} + \sigma \frac{\partial^2 u(t,x,y)}{\partial y^2}, \tag{10.1}$$

for $\Omega : (x,y) \in (0,1) \times (0,1)$ and for $t \leq T$. We let $\partial\Omega$ denote the boundary of the computational domain, $\Omega$. At the boundary, $\partial\Omega$, we use the Neumann boundary conditions

$$\frac{\partial u(t,0,y)}{\partial x} = 0, \quad \frac{\partial u(t,1,y)}{\partial x} = 0, \quad \frac{\partial u(t,x,0)}{\partial y} = 0, \quad \frac{\partial u(t,x,1)}{\partial y} = 0, \tag{10.2}$$

and, in addition, we define the initial condition

$$u(0,x,y) = u^0(x,y), \tag{10.3}$$

where $u^0(x,y)$ is a given function. In (10.1), $\sigma$ denotes a strictly positive (given) constant. In order to derive a numerical scheme for this problem, we proceed as usual by replacing derivatives by difference. By using the difference approximations introduced in Chapter 3, we get the following scheme[1],

$$\frac{u_{k,j}^{n+1} - u_{k,j}^n}{\Delta t} = \sigma \frac{u_{k-1,j}^n - 2u_{k,j}^n + u_{k+1,j}^n}{\Delta x^2} + \sigma \frac{u_{k,j-1}^n - 2u_{k,j}^n + u_{k,j+1}^n}{\Delta y^2}. \tag{10.4}$$

Here, $u_{k,j}^n$ denotes an approximation of $u(t_n,x_k,y_j)$ where $t_n = n\Delta t$, $x_k = (k-1)\Delta x$ and $y_j = (j-1)\Delta y$ with $\Delta x = 1/(M_x-1)$ and $\Delta y = 1/(M_y-1)$ for sufficiently large integers $M_x$ and $M_y$.

We noted above (see page 43) that it is quite useful to formulate these numerical schemes using matrix/vector notation. However, here the unknowns are given by $u_{k,j}^n$ and thus it is not straightforward to use such a notation. We therefore introduce a one-dimensional numbering of the two-dimensional problem. Specifically, we define the new index $i = M_x(j-1) + k$. Then, any two-dimensional vector $z$ with components $\{z_{k,j}\}$ can be rewritten as a one-dimensional vector $z_i$ where $i$ runs from 1 to $M = M_x \times M_y$. The components of the one-dimensional vector are given

---

[1] It may be useful to note the similarity with the one-dimensional case; see (3.13) at page 23.

**Fig. 10.1** Structure of the matrix $A$ used in the numerical scheme for the unitless diffusion equation. We have here used $\Delta x = \Delta y = 0.2$ and $\sigma = 0.04$, which gives $\rho_x = \rho_y = 1$, $M_x = M_y = 6$, and $M = M_x M_y = 36$.

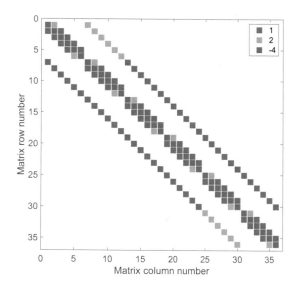

by $z_{M_x(j-1)+k}$ where $j$ and $k$ runs from 1 to $M_x$ and $M_y$, respectively. Using this numbering, we can rewrite the scheme (10.4) as follows,

$$\frac{u_i^{n+1} - u_i^n}{\Delta t} = \sigma \frac{u_{i-1}^n - 2u_i^n + u_{i+1}^n}{\Delta x^2} + \sigma \frac{u_{i-M_x}^n - 2u_i^n + u_{i+M_x}^n}{\Delta y^2}. \tag{10.5}$$

By defining $\rho_x = \sigma/\Delta x^2$ and $\rho_y = \sigma/\Delta y^2$, we can rewrite the scheme as follows,

$$u_i^{n+1} = u_i^n + \Delta t \left[ \rho_x(u_{i-1}^n + u_{i+1}^n) - 2(\rho_x + \rho_y)u_i^n + \rho_y(u_{i-M_x}^n + u_{i+M_x}^n) \right]. \tag{10.6}$$

Now, we can define an $M \times M$ matrix $A$ where non-zero elements of a typical row $i$ in the matrix are given by

$$a_{i,i-M_x} = \rho_y, \tag{10.7}$$

$$a_{i,i-1} = \rho_x, \tag{10.8}$$

$$a_{i,i} = -2(\rho_x + \rho_y), \tag{10.9}$$

$$a_{i,i+1} = \rho_x, \tag{10.10}$$

$$a_{i,i+M_x} = \rho_y. \tag{10.11}$$

When the index $i$ corresponds to a boundary point or to a corner point in the mesh, there are exceptions. The detailed definition of all the elements of the matrix is found in the software associated these notes, and a figure showing the structure of the matrix is depicted in Fig. 10.1 in the case $\Delta x = \Delta y = 0.2$ and $\sigma = 0.04$, which gives $\rho_x = \rho_y = 1$, $M_x = M_y = 6$, and $M = M_x M_y = 36$.

With this notation at hand, we can write the scheme (10.6) in the convenient form

$$u^{n+1} = (I + \Delta t A)u^n. \tag{10.12}$$

Furthermore, we can define the implicit scheme,

$$\frac{u^{n+1} - u^n}{\Delta t} = Au^{n+1}, \tag{10.13}$$

which can be rewritten as,

$$(I - \Delta t A)u^{n+1} = u^n, \tag{10.14}$$

where, as usual, $I$ denotes the identity matrix. We can also define the second order midpoint scheme as follows,

$$\frac{u^{n+1} - u^n}{\Delta t} = \frac{1}{2}A(u^n + u^{n+1}), \tag{10.15}$$

or

$$\left(I - \frac{\Delta t}{2}A\right)u^{n+1} = \left(I + \frac{\Delta t}{2}A\right)u^n. \tag{10.16}$$

## 10.2 The Bidomain Model

The bidomain model with membrane dynamics modeled by the parsimonious ventricular rabbit model [6] can be written in the form

$$\chi\left(C_m\frac{\partial v}{\partial t} + I_{\text{ion}}(v, m, h)\right) = \nabla \cdot (\sigma_i \nabla v) + \nabla \cdot (\sigma_i \nabla u_e), \tag{10.17}$$

$$0 = \nabla \cdot (\sigma_i \nabla v) + \nabla \cdot ((\sigma_i + \sigma_e)\nabla u_e), \tag{10.18}$$

$$\frac{dm}{dt} = \frac{m_\infty - m}{\tau_m}, \tag{10.19}$$

$$\frac{dh}{dt} = \frac{h_\infty - h}{\tau_h}. \tag{10.20}$$

We consider this system for $\Omega : (x, y) \in (0, L) \times (0, L)$ and for $t \leq T$. The spatial coordinate is given centimeters (cm) and time is given in milliseconds (ms). The unknown functions $v$ and $u_e$ are the membrane potential and the extracellular potential, respectively (in units of mV). Note that $v = u_i - u_e$, where $u_i$ is the intracellular potential. We can use any pair of variables among $v$, $u_i$ and $u_e$ as prime variables, but it is most common to use $v$ and $u_e$. The bidomain model can be derived by assuming that the membrane potential, $v$, the intracellular potential, $u_i$, and the extracellular potential, $u_e$, are all defined everywhere in the domain and following the same steps as those used to derive the cable equation (see Chapter 9) without the assumption of $u_e = 0$ (see Comment 2 in Section 9.4).

In the bidomain model, $\sigma_i$ and $\sigma_e$ denote the conductivities of the extracellular and intracellular spaces, respectively. The conductivities are tensors allowing the

**Table 10.1** Parameter values used in the bidomain model simulations.

| Parameter | Value |
|---|---|
| $C_m$ | $1\ \mu F/cm^2$ |
| $\sigma_{i,x}, \sigma_{i,y}$ | 3 mS/cm |
| $\sigma_{e,x}, \sigma_{e,y}$ | 10 mS/cm |
| $\chi$ | $2000\ cm^{-1}$ |
| $L_x, L_y$ | 1 cm |
| $\Delta x, \Delta y$ | 0.025 cm |
| $\Delta t$ | 0.01 ms |

conductivity to vary according to the spatial directions. Specifically, we have

$$\sigma_e = \begin{pmatrix} \sigma_{e,x} & 0 \\ 0 & \sigma_{e,y} \end{pmatrix}, \qquad \sigma_i = \begin{pmatrix} \sigma_{i,x} & 0 \\ 0 & \sigma_{i,y} \end{pmatrix}.$$

In general, the conductivities can vary in space, but here we will assume that they are constant. Furthermore, $C_m$ is the specific membrane capacitance, and $\chi$ denotes the surface-to-volume ratio of the cell membrane. The sum of the ion current densities across the membrane are given by

$$I_{ion}(v,m,h) = I_{Na}(v,m,h) + I_K(v) + I_{stim}. \tag{10.21}$$

The specific formulations for the current densities $I_{Na}$ and $I_K$ are as specified in Chapter 8 (see page 71), and so are the associated gating functions $m_\infty$, $h_\infty$, $\tau_m$ and $\tau_h$. The current density $I_{stim}$ represents a stimulus used to start the electrical wave. It is given by

$$I_{stim}(t,x,y) = \begin{cases} a_{stim}, & \text{if } t \geq t_{stim} \text{ and } t \leq t_{stim} + d_{stim}, \\ & \text{and } \sqrt{x^2+y^2} \leq l_{stim}, \\ 0, & \text{otherwise}, \end{cases} \tag{10.22}$$

where $a_{stim} = -25\ \mu A/cm^2$, $t_{stim} = 0$ ms, $d_{stim} = 2$ ms, and $l_{stim} = 0.25$ cm. The initial conditions are as specified for the parsimonious rabbit model in Chapter 8 (see page 72), and, for simplicity, we use the boundary conditions

$$u_e(t,0,y) = u_e(t,L_x,y) = u_e(t,x,0) = u_e(t,x,L_y) = 0, \tag{10.23}$$

$$\frac{\partial u_i(t,0,y)}{\partial x} = 0, \quad \frac{\partial u_i(t,L_x,y)}{\partial x} = 0, \quad \frac{\partial u_i(t,x,0)}{\partial y} = 0, \quad \frac{\partial u_i(t,x,L_y)}{\partial y} = 0. \tag{10.24}$$

### 10.2.1 Operator Splitting for the Bidomain Model

In order to solve the somewhat intimidating system (10.17)–(10.20) we use, more or less, all the tricks we have introduced above. The main technique, however, is to break the complex problem into parts that we are able to deal with. The first step along that path is to apply operator splitting. We start by assuming that the complete solution vector given by $(v, u_e, m, h)$ is known at time $t = t_n = n \times \Delta t$, and we want to compute the solution at time $t = t_{n+1}$. We do this in two steps. First we solve the following system of ordinary differential equations,

$$C_m \frac{\partial v}{\partial t} = -I_{\text{ion}}(v, m, h) \tag{10.25}$$

$$\frac{dm}{dt} = \frac{m_\infty - m}{\tau_m}, \tag{10.26}$$

$$\frac{dh}{dt} = \frac{h_\infty - h}{\tau_h}. \tag{10.27}$$

By solving these equations with $t$ ranging from $t_n$ to $t_{n+1}$, we obtain a solution that we denote $(v, u_e, m, h)^{n+1/2}$. In the next step, we solve the spatial part of the equation,

$$\chi C_m \frac{\partial v}{\partial t} = \nabla \cdot (\sigma_i \nabla v) + \nabla \cdot (\sigma_i \nabla u_e), \tag{10.28}$$

$$0 = \nabla \cdot (\sigma_i \nabla v) + \nabla \cdot ((\sigma_i + \sigma_e) \nabla u_e), \tag{10.29}$$

$$\frac{dm}{dt} = 0, \tag{10.30}$$

$$\frac{dh}{dt} = 0, \tag{10.31}$$

where the initial conditions at time $t = t_n$ is given by $(v, u_e, m, h)^{n+1/2}$. Here, we immediately note that $m^{n+1} = m^{n+1/2}$ and $h^{n+1} = h^{n+1/2}$. In order to compute $v$ and $u_e$ at $t_{n+1}$, we need to solve the linear system,

$$\chi C_m \frac{\partial v}{\partial t} = \nabla \cdot (\sigma_i \nabla v) + \nabla \cdot (\sigma_i \nabla u_e), \tag{10.32}$$

$$0 = \nabla \cdot (\sigma_i \nabla v) + \nabla \cdot ((\sigma_i + \sigma_e) \nabla u_e), \tag{10.33}$$

where the initial condition for $v$ is given by $v^{n+1/2}$.

### 10.2.2 Finite Difference Approximation

The task at hand is now to solve the system of ordinary differential equations given by (10.25)–(10.27) and then solve the partial differential equations given by

(10.32)-(10.33). In order to do this we introduce a mesh as above. Specifically[2], we let $u_{k,j}^n$ and $v_{k,j}^n$ denote approximations of $u(t_n, x_k, y_j)$ and $v(t_n, x_k, y_j)$ where $t_n = n\Delta t$, $x_k = (k-1)\Delta x$ and $y_j = (j-1)\Delta y$ with $\Delta x = L_x/(M_x - 1)$, $\Delta y = L_y/(M_y - 1)$ and $\Delta t = T/M_t$ for sufficiently large integers $M_x$, $M_y$ and $M_t$. Again, we use the mapping from a two-dimensional notation to a one-dimensional vector notation by defining $i = M_x(j-1) + k$.

## Explicit ODE Step

By using this notation, we can write an explicit scheme[3] for solving the ordinary differential equations given by (10.25)–(10.27) as follows,

$$v_i^{n+1/2} = v_i^n - \frac{\Delta t}{C_m}[I_{Na}(v_i^n, m_i^n, h_i^n) + I_K(v_i^n) + I_{stim}(t_n, x_i, y_i)], \qquad (10.34)$$

$$m_i^{n+1/2} = m_i^n + \Delta t \frac{m_\infty(v_i^n) - m_i^n}{\tau_m(v_i^n)}, \qquad (10.35)$$

$$h_i^{n+1/2} = h_i^n + \Delta t \frac{h_\infty(v_i^n) - h_i^n}{\tau_h(v_i^n)}, \qquad (10.36)$$

where $1 \le i \le M_x \times M_y$, and $n = 0, \ldots, M_t - 1$.

## Implicit PDE Step

In the implicit PDE step we first note that $m^{n+1} = m^{n+1/2}$ and $h^{n+1} = h^{n+1/2}$. For the systems of PDEs given by (10.32)–(10.33) we use the same discretization as we used in (10.4) for the diffusion equation. This leads to the definition of two matrices, $A_i$ and $A_e$, with typical elements given by (10.7)–(10.11). For $A_e$, the typical elements of the matrix are defined by using $\rho_x = \sigma_e/\Delta x^2$ and $\rho_y = \sigma_e/\Delta y^2$ in (10.7)–(10.11). Similarly, the typical elements of $A_i$ are defined by using $\rho_x = \sigma_i/\Delta x^2$ and $\rho_y = \sigma_i/\Delta y^2$ in (10.7)–(10.11). With this notation, we are ready to define the following scheme for the spatial part of the bidomain equations,

$$\chi C_m \frac{v^{n+1} - v^{n+1/2}}{\Delta t} = A_i v^{n+1} + A_i u^{n+1}, \qquad (10.37)$$

$$0 = A_i v^{n+1} + (A_i + A_e) u^{n+1}. \qquad (10.38)$$

This can be rewritten as a block matrix-vector system as follows,

---

[2] We replace $u_e$ by $u$ in order to reduce the load of subscripts.

[3] This is exactly the same scheme as we used for the parsimonious rabbit model in Section 8.2, only that we here need to solve the ODE system in all the computational nodes.

$$\begin{bmatrix} I - \frac{\Delta t}{\chi C_m} A_i & -\frac{\Delta t}{\chi C_m} A_i \\ A_i & A_i + A_e \end{bmatrix} \begin{bmatrix} v^{n+1} \\ u^{n+1} \end{bmatrix} = \begin{bmatrix} v^{n+1/2} \\ 0 \end{bmatrix}. \tag{10.39}$$

### 10.2.3  Traveling Wave Solution of the Bidomain Model

In Fig. 10.2, we show the solution of the numerical scheme for the bidomain model described above, with parameter values specified in Table 10.1 and membrane dynamics modeled by the parsimonious rabbit model described in Chapter 8. The upper panel shows the membrane potential, $v$, at four different points in time, and the lower panel shows the extracellular potential, $u_e$, at the same time points. We observe a traveling wave solution initiated by the stimulus current applied in the lower left corner of the domain, leading to an increased value of $v$. This increase in $v$ gradually spreads through the domain, and at $t = 20$ ms the wave has traveled almost all the way to the opposite corner of the domain.

From this traveling wave, we can compute the conduction velocity, for example, by

$$CV = \frac{\sqrt{(x_2 - x_1)^2 + (y_2 - y_1)^2}}{t_2 - t_1}, \tag{10.40}$$

where $x_1 = y_1 = 0.4$ cm and $x_2 = y_2 = 0.8$ cm. Furthermore, $t_1$ and $t_2$ are the points in time when the value of $v$ first increases to a value $v \geq -20$ mV, in the spatial points $(x_1, y_1)$ and $(x_2, y_2)$, respectively. In that case, we find that the conduction velocity in this default case is about 54 cm/s. In Fig. 10.3, we investigate how the conduction velocity depends on the conductivity parameters, $\sigma_i$ and $\sigma_e$, and we find that the conduction velocity increases if $\sigma_i$ or $\sigma_e$ are increased.

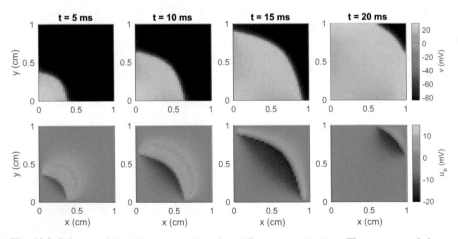

**Fig. 10.2** Solution of the bidomain model at four different points in time. The upper panel shows the membrane potential, $v$, and the lower panel shows the extracellular potential, $u_e$. A traveling wave solution is initiated by stimulating the cells in the lower left corner.

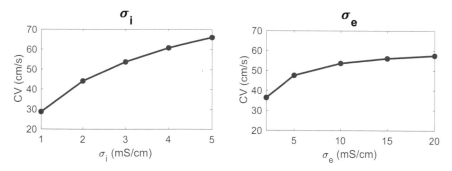

**Fig. 10.3** Conduction velocity computed from the solution of the bidomain model for some different values of $\sigma_i$ and $\sigma_e$. In the left plot, $\sigma_e$ is fixed at the value specified in Table 1.1, and in the right plot $\sigma_i$ is fixed at the value specified in Table 1.1.

## 10.3 The Monodomain Model

As mentioned above, the bidomain model is often referred to as the gold-standard for computational cardiac electrophysiology. But in many cases, results of relevant accuracy can be achieved by using the simpler monodomain model. Here will show that, under one specific condition, the two models actually give the same results. We can show this by considering the linear PDE that needs to be solved in each time step; see (10.32) and (10.33) above. We assume that the conductivities are constant in space, and that they are related as follows,

$$\begin{pmatrix} \sigma_{e,x} & 0 \\ 0 & \sigma_{e,y} \end{pmatrix} = \lambda \begin{pmatrix} \sigma_{i,x} & 0 \\ 0 & \sigma_{i,y} \end{pmatrix}, \tag{10.41}$$

where $\lambda$ is a positive constant. By using this assumption, we note that the second equation of the system

$$\chi C_m \frac{\partial v}{\partial t} = \nabla \cdot (\sigma_i \nabla v) + \nabla \cdot (\sigma_i \nabla u_e), \tag{10.42}$$

$$0 = \nabla \cdot (\sigma_i \nabla v) + \nabla \cdot ((\sigma_i + \sigma_e) \nabla u_e), \tag{10.43}$$

can be rewritten as follows,

$$0 = \nabla \cdot (\sigma_i \nabla v) + (1 + \lambda) \nabla \cdot (\sigma_i \nabla u_e), \tag{10.44}$$

and therefore,

$$\nabla \cdot (\sigma_i \nabla u_e) = -\frac{1}{1 + \lambda} \nabla \cdot (\sigma_i \nabla v). \tag{10.45}$$

By inserting this observation in (10.42), we observe that we can remove $u_e$ from this equation and get the following scalar equation (only $v$ is unknown here),

$$\chi C_m \frac{\partial v}{\partial t} = \frac{\lambda}{1+\lambda} \nabla \cdot (\sigma_i \nabla v). \tag{10.46}$$

### 10.3.1  Operator Splitting for the Monodomain System

In the case of the bidomain model, operator splitting involved alternating solution of the ODEs by the scheme (10.34)–(10.36) and the PDEs by solving the linear system (10.37) and (10.38). For the monodomain case, the PDE part can be simplified considerably by solving the system

$$\chi C_m \frac{v^{n+1} - v^{n+1/2}}{\Delta t} = \frac{\lambda}{1+\lambda} A_i v^{n+1}, \tag{10.47}$$

$$\tag{10.48}$$

which can be written to the computational form

$$(I - \gamma \Delta t A_i) v^{n+1} = v^{n+1/2}, \tag{10.49}$$

$$\tag{10.50}$$

where

$$\gamma = \frac{\lambda}{(1+\lambda)\chi C_m}.$$

## 10.4  Comments and Further Reading

1. Introductions to the bidomain and monodomain models can be found in, e.g., [4, 16].
2. Numerical simulations of cardiac electrophysiology based on the bidomain and monodomain equations have been applied in a very large number of papers. Since 1990, a steady stream of results have been produced by many excellent research groups. There are far too many studies for us to review here; instead we will refer you to a few papers that have become classics in the field; [3, 5, 12, 13, 18, 19].
3. Numerical solution of the Poisson equation is one of the best studied problems in scientific computing. The theory of fast linear solvers is very well developed for symmetric and positive definite linear systems. The linear systems arising from the bidomain model represent extensions of the classical systems generated from the Poisson equation and have therefore received substantial interest. In the system (10.39), the matrices $A_i$ and $A_e$ are symmetric and thus the complete system can be written in a symmetric form if the lower part of the system is multiplied by $-\frac{\Delta t}{\chi C_m}$. Then the system reads

$$
\begin{bmatrix} I - \frac{\Delta t}{\chi C_m} A_i & -\frac{\Delta t}{\chi C_m} A_i \\ -\frac{\Delta t}{\chi C_m} A_i & -\frac{\Delta t}{\chi C_m}(A_i + A_e) \end{bmatrix} \begin{bmatrix} v^{n+1} \\ u^{n+1} \end{bmatrix} = \begin{bmatrix} v^{n+1/2} \\ 0 \end{bmatrix}. \tag{10.51}
$$

This system is symmetric and positive definite and therefore fast solvers can be applied; see, e.g., [2, 4, 7, 8, 10, 11, 14, 15, 16, 17].

4. The units used in spatially resolved electrophysiology models can be horrible to keep track of and it is not uncommon to miss the target by a factor of $10^3$ or even $10^6$, and little sense arises from such computations. After some time you will get used to this difficulty and will at least become wise enough to hide your blunders until they are properly corrected. As usual, there is no way around this but blood, toil, sweat and tears. However, finite differences can actually be of some assistance. We find it easier to check the units when a system of equations is written in the form of a finite difference scheme, because the derivatives in the original equations often cause confusion, and differences are easier to deal with. Note also that the basic rules are quite simple; all terms in a sum must be expressed in terms of the same unit, and so must the left- and right-hand sides of an equation.

5. There are two main reasons for approximating the bidomain model by the monodomain model. The first reason is the complexity of solving the equations. The monodomain model is of a very classical form and software are available from other fields and in general finite element libraries. The bidomain model, on the other hand, is less standard and implementing solution methods is therefore considered more challenging. Note, however, that excellent open-source software libraries are available; see, e.g., https://opencarp.org. The second reason is that the computational complexity (CPU-efforts needed to solve the equations) is often regarded to be much higher for the bidomain model than for the the monodomain model. This strongly depends on the methods used to solve the equations; see, e.g., [17].

6. It has been shown that solutions of the monodomain model, in many cases, provide very good approximations of the solutions of the bidomain model, see, e.g., [12].

7. The bidomain model is often hailed as the gold standard of computational electrophysiology, and we have added to this acclaim. Forty years ago, it was almost unthinkable to perform simulations of a whole heart because of the computational complexity. In 1984, it was estimated that it would take 3000 years to solve the bidomain model for 10 ms using a mesh with a million nodes [1], whereas the representation of the full human heart required about 26 million nodes. This estimate was clearly on the pessimistic side, since in 2006, the full simulation was performed in only two days (see [12]), and a few years later such simulations could be performed in minutes (see, e.g., [9]). Today, the bidomain model is used routinely and simulation times are acceptable even without extreme computing facilities.

# References

[1] Barr RC, Plonsey R (1984) Propagation of excitation in idealized anisotropic two-dimensional tissue. Biophysical Journal 45(6):1191–1202

[2] Del Corso G, Verzicco R, Viola F (2022) A fast computational model for the electrophysiology of the whole human heart. Journal of Computational Physics 457:111084

[3] Franzone PC, Pavarino L, Taccardi B (2005) Simulating patterns of excitation, repolarization and action potential duration with cardiac bidomain and monodomain models. Mathematical Biosciences 197(1):35–66

[4] Franzone PC, Pavarino LF, Scacchi S (2014) Mathematical Cardiac Electrophysiology, vol 13. Springer

[5] Geselowitz DB, Miller W (1983) A bidomain model for anisotropic cardiac muscle. Annals of Biomedical Engineering 11(3):191–206

[6] Gray RA, Pathmanathan P (2016) A parsimonious model of the rabbit action potential elucidates the minimal physiological requirements for alternans and spiral wave breakup. PLoS Computational Biology 12(10):e1005087

[7] Mardal KA, Winther R (2011) Preconditioning discretizations of systems of partial differential equations. Numerical Linear Algebra with Applications 18(1):1–40

[8] Mardal KA, Nielsen BF, Cai X, Tveito A (2007) An order optimal solver for the discretized bidomain equations. Numerical Linear Algebra with Applications 14(2):83–98

[9] Niederer S, Mitchell L, Smith N, Plank G (2011) Simulating human cardiac electrophysiology on clinical time-scales. Frontiers in Physiology 2:14

[10] Pavarino LF, Scacchi S (2008) Multilevel additive Schwarz preconditioners for the Bidomain reaction-diffusion system. SIAM Journal on Scientific Computing 31(1):420–443

[11] Plank G, Liebmann M, dos Santos RW, Vigmond EJ, Haase G (2007) Algebraic multigrid preconditioner for the cardiac bidomain model. IEEE Transactions on Biomedical Engineering 54(4):585–596

[12] Potse M, Dubé B, Richer J, Vinet A, Gulrajani RM (2006) A comparison of monodomain and bidomain reaction-diffusion models for action potential propagation in the human heart. IEEE Transactions on Biomedical Engineering 53(12):2425–2435

[13] Roth BJ (1997) Electrical conductivity values used with the bidomain model of cardiac tissue. IEEE Transactions on Biomedical Engineering 44(4):326–328

[14] dos Santos RW, Plank G, Bauer S, Vigmond EJ (2004) Parallel multigrid preconditioner for the cardiac bidomain model. IEEE Transactions on Biomedical Engineering 51(11):1960–1968

[15] Sundnes J, Lines G, Mardal K, Tveito A (2002) Multigrid block preconditioning for a coupled system of partial differential equations modeling the electrical activity in the heart. Computer Methods in Biomechanics & Biomedical Engineering 5(6):397–409

[16] Sundnes J, Lines GT, Cai X, Nielsen BF, Mardal KA, Tveito A (2006) Computing the electrical activity in the heart. Springer

[17] Sundnes J, Nielsen BF, Mardal KA, Cai X, Lines GT, Tveito A (2006) On the computational complexity of the bidomain and the monodomain models of electrophysiology. Annals of Biomedical Engineering 34(7):1088–1097

[18] Vigmond E, Dos Santos RW, Prassl A, Deo M, Plank G (2008) Solvers for the cardiac bidomain equations. Progress in Biophysics and Molecular Biology 96(1-3):3–18

[19] Vigmond EJ, Aguel F, Trayanova NA (2002) Computational techniques for solving the bidomain equations in three dimensions. IEEE Transactions on Biomedical Engineering 49(11):1260–1269

# Chapter 11
# Re-Introducing the Cell: The Extracellular-Membrane-Intracellular (EMI) Model

As mentioned earlier, the bidomain system is currently the standard mathematical model of cardiac electrophysiology. This system is now routinely solved and provides valuable insights into the conduction of electrical signals in cardiac tissue. However, the model has one glaring limitation: The cardiomyocyte is nowhere to be found in the model, since the extracellular space, the intracellular space and the cell membrane are all assumed to be everywhere in the computational domain. The cell was lost in homogenization! There is a tremendous advantage to this because the model becomes much simpler and thus solvable for the whole human heart. And it works! But the downside is of course that the cell is the essential building block of the tissue and leaving it out of the model has consequences. For instance, it becomes impossible to investigate the detailed dynamics of the electrochemical processes in the vicinity of a small collection of cells.

Here, we will present an alternative cell-based model. The model represents the extracellular (E) domain, the cell membrane (M) and the intracellular (I) domain explicitly and it is therefore referred to as the EMI model. The main advantage of this model is that it becomes feasible to represent the cell and the cell membrane in a much more detailed manner. For instance, it is possible to study both the effect of varying ion channel density across the cell membrane and cell to cell variations of properties in the model. But it comes with a stiff price: both implementation and computing efforts are much more demanding than for the bidomain model.

Here, we will present the EMI model and observe that, again, we can come up with a reasonable solution method by applying operator splitting and replacing derivatives with differences. We will present a case that is as simple as possible but also give references to more challenging applications.

© The Author(s) 2023
K. Horgmo Jæger, A. Tveito, *Differential Equations for Studies in Computational Electrophysiology*,
Simula SpringerBriefs on Computing 14, https://doi.org/10.1007/978-3-031-30852-9_11

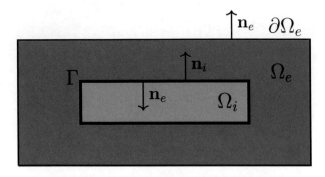

**Fig. 11.1** Illustration of the different parts of the domain in the EMI model. A cell, $\Omega_i$ is surrounded by an extracellular space, $\Omega_e$. The interface between $\Omega_i$ and $\Omega_e$ defines the cell membrane, $\Gamma$. Note that all the EMI model simulations are performed in 3D.

## 11.1 The EMI Model for one Cell Surrounded by an Extracellular Space

The system of partial differential equations forming the EMI model for a single cell surrounded by an extracellular space like illustrated in Fig. 11.1 is given by (see, e.g., [1, 2, 14, 41]):

$$\nabla \cdot \sigma_i \nabla u_i = 0, \qquad\qquad \text{in } \Omega_i, \qquad (11.1)$$

$$\nabla \cdot \sigma_e \nabla u_e = 0, \qquad\qquad \text{in } \Omega_e, \qquad (11.2)$$

$$u_e = 0, \qquad\qquad \text{at } \partial\Omega_e, \qquad (11.3)$$

$$\mathbf{n}_e \cdot \sigma_e \nabla u_e = -\mathbf{n}_i \cdot \sigma_i \nabla u_i, \qquad\qquad \text{at } \Gamma, \qquad (11.4)$$

$$u_i - u_e = v, \qquad\qquad \text{at } \Gamma, \qquad (11.5)$$

$$I_m = -\mathbf{n}_i \cdot \sigma_i \nabla u_i, \qquad\qquad \text{at } \Gamma, \qquad (11.6)$$

$$\frac{\partial v}{\partial t} = \frac{1}{C_m}(I_m - I_{\text{ion}}), \qquad\qquad \text{at } \Gamma. \qquad (11.7)$$

Here, the unknown variables to be found are the intracellular potential, $u_i$, the extracellular potential, $u_e$, and the membrane potential, $v$, all given in units of millivolts (mV). The intracellular potential is defined in the intracellular space, $\Omega_i$, the extracellular potential is defined in the extracellular space, $\Omega_e$, and the membrane potential is defined at the membrane, $\Gamma$, defined as the interface between $\Omega_i$ and $\Omega_e$. The outer boundary of the extracellular space is denoted by $\partial\Omega_e$. Time is given in milliseconds (ms) and distance is specified in centimeters (cm). Furthermore, $\sigma_i$ is the intracellular conductivity (in mS/cm), $\sigma_e$ is the extracellular conductivity (in mS/cm), $C_m$ is the specific membrane capacitance (in $\mu$S/cm$^2$), and $\mathbf{n}_i$ and $\mathbf{n}_e$ are the outward pointing unit normal vectors of $\Omega_i$ and $\Omega_e$, respectively. Like in the membrane models in Chapter 8, the cable equation in Chapter 9 and in the bidomain and monodomain models in Chapter 10, the term $I_{\text{ion}}$ represents the current density (in $\mu$A/cm$^2$) through ion channels on the cell membrane. These current densities can, for example, be modeled by the Hodgkin-Huxley model described in Chapter 8. In that case,

$$I_{\text{ion}} = I_{\text{Na}} + I_{\text{K}} + I_{\text{L}}, \tag{11.8}$$

and we get the following extra equations for the state variables defined at the membrane, $\Gamma$:

$$\frac{\partial m}{\partial t} = \alpha_m(1-m) - \beta_m m, \qquad \text{at } \Gamma, \tag{11.9}$$

$$\frac{\partial h}{\partial t} = \alpha_h(1-h) - \beta_h h, \qquad \text{at } \Gamma, \tag{11.10}$$

$$\frac{\partial r}{\partial t} = \alpha_r(1-r) - \beta_r r, \qquad \text{at } \Gamma. \tag{11.11}$$

### 11.1.1 Numerical Scheme for the EMI Model

As observed in the previous chapters, a numerical finite difference scheme for the EMI model can be defined by applying operator splitting and replacing derivatives with differences. We define a scheme where for each time step, $n$, there is one unknown, $u_e^n$, for all mesh points in the extracellular space and one unknown, $u_i^n$, for each intracellular point. In addition, for the mesh points located on the membrane, there are six unknowns, $u_i^n$, $u_e^n$, $v^n$, $m^n$, $h^n$, and $r^n$.

#### Operator Splitting for the EMI Model

We define an operator splitting scheme for the EMI model where for each time step we first solve the nonlinear ordinary differential equation part of the system

$$\frac{\partial v}{\partial t} = -\frac{1}{C_m} I_{\text{ion}}, \qquad \text{at } \Gamma, \tag{11.12}$$

$$\frac{\partial m}{\partial t} = \alpha_m(1-m) - \beta_m m, \qquad \text{at } \Gamma, \tag{11.13}$$

$$\frac{\partial h}{\partial t} = \alpha_h(1-h) - \beta_h h, \qquad \text{at } \Gamma, \tag{11.14}$$

$$\frac{\partial r}{\partial t} = \alpha_r(1-r) - \beta_r r, \qquad \text{at } \Gamma, \tag{11.15}$$

with initial conditions from the previous time step. Then, in the second step of the operator splitting scheme, we solve the linear system,

$$\nabla \cdot \sigma_i \nabla u_i = 0, \qquad\qquad \text{in } \Omega_i, \qquad (11.16)$$

$$\nabla \cdot \sigma_e \nabla u_e = 0, \qquad\qquad \text{in } \Omega_e, \qquad (11.17)$$

$$u_e = 0, \qquad\qquad \text{at } \partial\Omega_e, \qquad (11.18)$$

$$\mathbf{n}_e \cdot \sigma_e \nabla u_e = -\mathbf{n}_i \cdot \sigma_i \nabla u_i, \qquad \text{at } \Gamma, \qquad (11.19)$$

$$u_i - u_e = v, \qquad\qquad \text{at } \Gamma, \qquad (11.20)$$

$$I_m = -\mathbf{n}_i \cdot \sigma_i \nabla u_i, \qquad \text{at } \Gamma, \qquad (11.21)$$

$$\frac{\partial v}{\partial t} = \frac{1}{C_m} I_m, \qquad\qquad \text{at } \Gamma, \qquad (11.22)$$

with initial conditions provided from the first step of the operator splitting scheme.

**Finite Difference Approximation of the EMI Model**

The first step of the operator splitting scheme for the EMI model is simply a system of ordinary differential equations, and we find the numerical solutions by replacing the derivatives in the form $\frac{\partial v}{\partial t}$ with the differences in the form $\frac{v^{n+1}-v^n}{\Delta t}$ in an explicit manner.

In the second step of the operator splitting scheme, we also replace the temporal derivative in (11.22) with the standard difference, but we here treat the system in an implicit manner. That is, we replace (11.22) with

$$\frac{v^{n+1} - v^n}{\Delta t} = \frac{1}{C_m} I_m^{n+1}. \qquad (11.23)$$

Furthermore, we use standard differences for the derivatives in (11.16), (11.17), (11.19), and (11.21). However, some special treatment is required for the normal derivatives in (11.19) and (11.21) at the corners of the cell. The details of the finite difference scheme is found in the code associated with these notes and is also described in more detail in [41].

## 11.1.2 EMI Model Simulation of a Neuronal Axon

Using the parameter values specified in Table 11.1, we perform an EMI model simulation of a neuronal axon using a similar setup as for the cable equation in Chapter 9. However, to reduce the computational cost, we consider a shorter axon than in Chapter 9 (0.2 cm). A traveling wave moving from left to right is initiated by increasing the membrane potential of the leftmost 0.05 cm of the axon. Fig. 11.2 shows the numerical EMI model solution of the problem at three different points in time along a plane in the $x$- and $y$-directions. The left panel shows the extracellular potential, and the right panel shows the membrane potential, $v$.

**Table 11.1** Parameter values used in the EMI model simulations of an axon. The parameter values of the Hodgkin-Huxley model are as specified in Chapter 8.

| Parameter | Value | Parameter | Value |
|---|---|---|---|
| $C_m$ | $1 \, \mu\text{F/cm}^2$ | $\Omega_i$ | $2000 \, \mu\text{m} \times 10 \, \mu\text{m} \times 10 \, \mu\text{m}$ |
| $\sigma_i$ | $4 \, \text{mS/cm}$ | $\Omega_i \cup \Omega_e$ | $2060 \, \mu\text{m} \times 50 \, \mu\text{m} \times 50 \, \mu\text{m}$ |
| $\sigma_e$ | $3 \, \text{mS/cm}$ | $\Delta x$ | $10 \, \mu\text{m}$ |
| $\Delta t$ | $0.02 \, \text{ms}$ | $\Delta y, \Delta z$ | $2.5 \, \mu\text{m}$ |

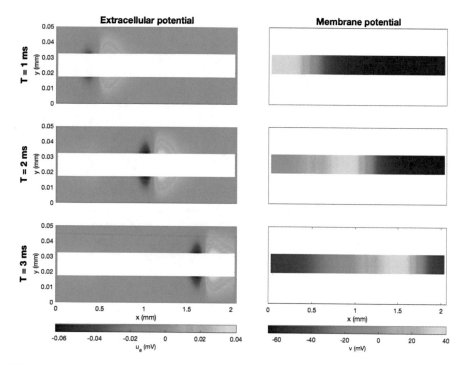

**Fig. 11.2** EMI model solution for three points in time along a plane in the $x$- and $y$-directions. The plane is located at the $z$-value corresponding to the upper boundary of the cell. The left panel shows the extracellular potential, and the right panel shows the membrane potential, $v$.

## 11.2 The EMI Model for Connected Cardiomyocytes

The EMI model for one cell considered in the previous section can be extended to incorporate currents between individual cells through gap junctions and thus be used to model collections of connected cardiomyocytes (see, e.g., [14, 16, 29, 35, 36, 37]). The resulting system of equations can be solved using a similar operator splitting technique as the one applied for a single cell above (see, e.g., [40]). Furthermore, a spatial operator splitting approach can be introduced in order to split the linear part

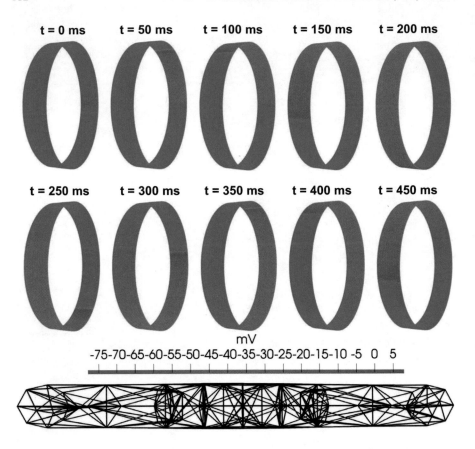

**Fig. 11.3** EMI model solution (intracellular potential, $u_i$) for a pulmonary vein sleeve at ten points in time, following the simulation set-up used in [20]. The two mutations N588K and A130V are both present. The solutions are found using the finite element method (see [20]). The finite element mesh used to represent each single cardiomyocyte in the simulation is illustrated in the lower panel of the figure. The full cylinder of cells seen in the upper panels contains 3930 cardiomyocytes, each associated with about 70 computational nodes. In addition, the mesh consists of about 44,000 extracellular nodes.

of the system system into one system for the extracellular space and one system for each individual cell (see [17, 19]).

## 11.2.1  EMI Model Simulation of Cardiomyocytes in the Sleeve of a Pulmonary Vein

To illustrate an example of the EMI model used for connected cardiomyocytes, we consider a collection of myocytes located around the sleeve of a pulmonary vein. In [20], the EMI model was used to study how mutations that have been found to be associated with increased risk of atrial fibrillation affected properties of the cardiomyocytes in this region, known to be a common initiation site of atrial fibrillation. In Fig. 11.3, we consider a collection of cells forming a cylinder around the vein, using the same setup as in [20]. The mesh used to represent each single cell in the simulation is illustrated in the lower panel of the figure. The properties of the individual cells vary according to known differences for cardiomyocytes in this region (see [20]). In the simulation, a combination of two mutations found to be associated with atrial fibrillation is present. The first mutation, N588K, leads to an increased potassium current ($I_{Kr}$), and thus shortening of the action potential duration, whereas the second mutation, A130V, leads to reduced sodium current ($I_{Na}$) and thereby reduced conduction velocity. In Fig. 11.3, we observe that when the two mutations are present, the solution is a traveling wave continuously moving around the cylinder of cells. Such a reentrant excitation wave could be a potential mechanism of atrial fibrillation.

## 11.3 Comments and Further Reading

1. As mentioned above, the EMI model can be used to represent the individual cells of cardiac tissue and can therefore be referred to as a cell-based model, as opposed to the homogenized bidomain and monodomain models. Alternative cell-based models have also been introduced, including 1D single strand models (e.g., [6, 22, 30, 39, 42]), 2D sheet models (e.g., [9, 10, 11, 12, 21, 31, 32, 33, 34]), and 3D microdomain models (e.g., [24, 25, 26]). Differences between the EMI model and other cell-based models are discussed in [16].
2. The resolution used in monodomain and bidomain model simulations is often about $\Delta x \approx 0.25$ mm (see, e.g., [5, 43]). The volume of a cardiomyocyte has been reported to be around 16 pL, see [28]. Every computational block with volume $(0.25 \text{ mm})^3$ can thus cover almost 1000 cardiomyocytes (see [18]). This means that homogenization is very efficient in removing lots of details, which is good for computing efforts. But it also means that lots of details are lost, which is bad news for understanding the physics at the level of the cells.
3. As mentioned above, the time to solve the bidomain model for a million nodes was estimated to 3000 years in 1984. So what is the estimate for EMI (now, in 2023)? According to [18] the computing time for one time step is about 0.02 ms for each cell. For an action potential lasting for 500 ms, the total number of time steps is $500 \times 10^3$ when the time step is $\Delta t = 0.001$ ms. The computing time per cell for an action potential of 500 ms is therefore about 10 seconds. This means that we can easily deal with small collections of cells. Simulating 1000 cells would take less than three hours. But the human heart contains between 2 and 3 billion cells (see [38]). The computing time for 2 billion cells for one action potential is about $2^9 \times 10$ seconds or 23,148 days or about 63 years. So it is not as bad as the bidomain model anno 1984, but it is a long wait!

4. What is the relation between the EMI model and the bidomain model? The bidomain model was developed long ago and can be derived in many different ways, but it can also be directly obtain by averaging the EMI equations over many cells (see [15]).
5. Derivations of the EMI model can be found in [1, 13, 14]. The EMI has been used to study a number of different electrophysiological phenomena, including applications relevant for both neuroscience (e.g., [2, 3, 4, 41]) and cardiac electrophysiology (e.g., [16, 20, 29, 35, 36, 37]). Furthermore, a number of numerical strategies for solving the equations have been proposed, including both finite difference and finite element schemes (see, e.g., [1, 17, 19, 23, 40]).
6. One of the simplifying assumptions underlying the EMI model presented in this chapter is that the effect of diffusion of ions in the intracellular and extracellular spaces are ignored. Such diffusion effects can be included in the model by including the Kirchhoff-Nernst-Planck (KNP) equations in the model, sometimes referred to as KNP-EMI models (see, e.g., [7, 8, 27]).

# References

[1] Agudelo-Toro A (2012) Numerical simulations on the biophysical foundations of the neuronal extracellular space. PhD thesis, Niedersächsische Staats-und Universitätsbibliothek Göttingen
[2] Agudelo-Toro A, Neef A (2013) Computationally efficient simulation of electrical activity at cell membranes interacting with self-generated and externally imposed electric fields. Journal of Neural Engineering 10(2):026019
[3] Buccino AP, Kuchta M, Jæger KH, Ness TV, Berthet P, Mardal KA, Cauwenberghs G, Tveito A (2019) How does the presence of neural probes affect extracellular potentials? Journal of Neural Engineering 16(2):026030
[4] Buccino AP, Kuchta M, Schreiner J, Mardal KA (2021) Improving neural simulations with the EMI model. In: Modeling Excitable Tissue, Springer, pp 87–98
[5] Clayton R, Panfilov A (2008) A guide to modelling cardiac electrical activity in anatomically detailed ventricles. Progress in Biophysics and Molecular Biology 96(1-3):19–43
[6] Copene ED, Keener JP (2008) Ephaptic coupling of cardiac cells through the junctional electric potential. Journal of Mathematical Biology 57(2):265–284
[7] Ellingsrud AJ, Solbrå A, Einevoll GT, Halnes G, Rognes ME (2020) Finite element simulation of ionic electrodiffusion in cellular geometries. Frontiers in Neuroinformatics 14:11
[8] Ellingsrud AJ, Daversin-Catty C, Rognes ME (2021) A cell-based model for ionic electrodiffusion in excitable tissue. In: Modeling Excitable Tissue, Springer, pp 14–27

[9] Hubbard ML, Henriquez CS (2010) Increased interstitial loading reduces the effect of microstructural variations in cardiac tissue. American Journal of Physiology-Heart and Circulatory Physiology 298(4):H1209–H1218

[10] Hubbard ML, Henriquez CS (2012) Microscopic variations in interstitial and intracellular structure modulate the distribution of conduction delays and block in cardiac tissue with source–load mismatch. Europace 14:v3–v9

[11] Hubbard ML, Henriquez CS (2014) A microstructural model of reentry arising from focal breakthrough at sites of source-load mismatch in a central region of slow conduction. American Journal of Physiology-Heart and Circulatory Physiology 306(9):H1341–H1352

[12] Hubbard ML, Ying W, Henriquez CS (2007) Effect of gap junction distribution on impulse propagation in a monolayer of myocytes: a model study. Europace 9:vi20–vi28

[13] Jæger KH (2019) Cell-based mathematical models of small collections of excitable cells. PhD thesis, University of Oslo

[14] Jæger KH, Tveito A (2021) Derivation of a cell-based mathematical model of excitable cells. In: Modeling Excitable Tissue, Springer, pp 1–13

[15] Jæger KH, Tveito A (2022) Deriving the bidomain model of cardiac electrophysiology from a cell-based model; properties and comparisons. Frontiers in Physiology 12:811029

[16] Jæger KH, Edwards AG, McCulloch A, Tveito A (2019) Properties of cardiac conduction in a cell-based computational model. PLoS Computational Biology 15(5):e1007042

[17] Jæger KH, Hustad KG, Cai X, Tveito A (2020) Operator splitting and finite difference schemes for solving the EMI model. Modeling Excitable Tissue pp 44–55

[18] Jæger KH, Edwards AG, Giles WR, Tveito A (2021) From millimeters to micrometers; re-introducing myocytes in models of cardiac electrophysiology. Frontiers in Physiology 12:763584

[19] Jæger KH, Hustad KG, Cai X, Tveito A (2021) Efficient numerical solution of the EMI model representing the extracellular space (E), cell membrane (M) and intracellular space (I) of a collection of cardiac cells. Frontiers in Physics 8:579461

[20] Jæger KH, Edwards AG, Giles WR, Tveito A (2022) Arrhythmogenic influence of mutations in a myocyte-based computational model of the pulmonary vein sleeve. Scientific Reports 12(1):1–18

[21] Joyner RW, Ramón F, Morre J (1975) Simulation of action potential propagation in an inhomogeneous sheet of coupled excitable cells. Circulation Research 36(5):654–661

[22] Kucera JP, Rohr S, Rudy Y (2002) Localization of sodium channels in intercalated disks modulates cardiac conduction. Circulation Research 91(12):1176–1182

[23] Kuchta M, Mardal KA, Rognes ME (2021) Solving the EMI equations using finite element methods. In: Modeling Excitable Tissue, Springer, pp 56–69

[24] Lin J, Keener JP (2010) Modeling electrical activity of myocardial cells incorporating the effects of ephaptic coupling. Proceedings of the National Academy of Sciences 107(49):20935–20940

[25] Lin J, Keener JP (2013) Ephaptic coupling in cardiac myocytes. IEEE Transactions on Biomedical Engineering 60(2):576–582

[26] Lin J, Keener JP (2014) Microdomain effects on transverse cardiac propagation. Biophysical Journal 106(4):925–931

[27] Mori Y, Fishman GI, Peskin CS (2008) Ephaptic conduction in a cardiac strand model with 3D electrodiffusion. Proceedings of the National Academy of Sciences 105(17):6463–6468

[28] Nygren A, Fiset C, Firek L, Clark JW, Lindblad DS, Clark RB, Giles WR (1998) Mathematical model of an adult human atrial cell: the role of K+ currents in repolarization. Circulation Research 82(1):63–81

[29] Roberts SF, Stinstra JG, Henriquez CS (2008) Effect of nonuniform interstitial space properties on impulse propagation: a discrete multidomain model. Biophysical Journal 95(8):3724–3737

[30] Shaw RM, Rudy Y (1997) Ionic mechanisms of propagation in cardiac tissue: roles of the sodium and L-type calcium currents during reduced excitability and decreased gap junction coupling. Circulation Research 81(5):727–741

[31] Spach MS, Heidlage JF (1995) The stochastic nature of cardiac propagation at a microscopic level: electrical description of myocardial architecture and its application to conduction. Circulation Research 76(3):366–380

[32] Spach MS, Heidlage JF, Dolber PC, Barr RC (1998) Extracellular discontinuities in cardiac muscle: evidence for capillary effects on the action potential foot. Circulation Research 83(11):1144–1164

[33] Spach MS, Heidlage JF, Dolber PC, Barr RC (2000) Electrophysiological effects of remodeling cardiac gap junctions and cell size. Circulation Research 86(3):302–311

[34] Spach MS, Heidlage JF, Dolber PC, Roger C (2001) Changes in anisotropic conduction caused by remodeling cell size and the cellular distribution of gap junctions and Na+ channels. Journal of Electrocardiology 34(4):69–76

[35] Stinstra J, MacLeod R, Henriquez C (2010) Incorporating histology into a 3D microscopic computer model of myocardium to study propagation at a cellular level. Annals of Biomedical Engineering 38(4):1399–1414

[36] Stinstra JG, Hopenfeld B, MacLeod RS (2005) On the passive cardiac conductivity. Annals of Biomedical Engineering 33(12):1743–1751

[37] Stinstra JG, Roberts SF, Pormann JB, MacLeod RS, Henriquez CS (2006) A model of 3D propagation in discrete cardiac tissue. In: Computers in Cardiology, 2006, IEEE, vol 33, pp 41–44

[38] Tirziu D, Giordano FJ, Simons M (2010) Cell communications in the heart. Circulation 122(9):928–937

[39] Tsumoto K, Ashihara T, Haraguchi R, Nakazawa K, Kurachi Y (2011) Roles of subcellular Na+ channel distributions in the mechanism of cardiac conduction. Biophysical Journal 100(3):554–563

[40] Tveito A, Jæger KH, Kuchta M, Mardal KA, Rognes ME (2017) A cell-based framework for numerical modeling of electrical conduction in cardiac tissue. Frontiers in Physics 5:48

[41] Tveito A, Jæger KH, Lines GT, Paszkowski Ł, Sundnes J, Edwards AG, Māki-Marttunen T, Halnes G, Einevoll GT (2017) An evaluation of the accuracy of classical models for computing the membrane potential and extracellular potential for neurons. Frontiers in Computational Neuroscience 11:27

[42] Wang Y, Rudy Y (2000) Action potential propagation in inhomogeneous cardiac tissue: safety factor considerations and ionic mechanism. American Journal of Physiology-Heart and Circulatory Physiology 278(4):H1019–H1029

[43] Xie F, Qu Z, Yang J, Baher A, Weiss JN, Garfinkel A, et al. (2004) A simulation study of the effects of cardiac anatomy in ventricular fibrillation. The Journal of Clinical Investigation 113(5):686–693

# Chapter 12
# The Poisson-Nernst-Planck (PNP) Model

In these notes, we have considered models of electrophysiology across several scales. The first was the membrane model. It assumes that the action potential is similar across the whole cell membrane, and the model represents the action potential as a function of time alone. No spatial variable is involved in the pure membrane models, so a length scale of these models does not make sense. As models of cardiac electrophysiology, we next considered the monodomain and bidomain equations. These models are accurate descriptions of the physics at the scale of millimeters, and the typical mesh resolution is about $\Delta x \approx 0.25$ mm (see, e.g., [1, 16]). The EMI model is cell-based and represents the physics at the micrometer scale. The typical mesh size for the EMI model is about 10 $\mu$m see, e.g., [7, 8, 9].

We have moved from the homogenized millimeter scale (monodomain/bidomain) to the cell-based EMI model on the micrometer scale. Next, we move to the nanometer scale. The reason for this is that strong electrical and chemical gradients exist very close to the cell membrane. These gradients, referred to as the Debye layer (see, e.g., [6]), are only a few nanometers wide. In order to study what happens very close to the membrane, it is necessary to solve equations on the nanometer level, and the proper equations are referred to as the Poisson-Nernst-Planck (PNP) model. Once again we will see that even if the model is rather complex, we can use the standard tricks introduced above to solve the equations; operator splitting and finite differences are all we need (plus a little blood, toil, sweat and tears).

© The Author(s) 2023
K. Horgmo Jæger, A. Tveito, *Differential Equations for Studies in Computational Electrophysiology*,
Simula SpringerBriefs on Computing 14, https://doi.org/10.1007/978-3-031-30852-9_12

## 12.1  The PNP System of Partial Differential Equations

The PNP system can be written in the form

$$\nabla \cdot (\varepsilon \nabla \phi) = -\rho, \tag{12.1}$$

$$\frac{\partial c_k}{\partial t} = \nabla \cdot D_k \nabla c_k + \nabla \cdot (D_k \beta_k c_k \nabla \phi), \tag{12.2}$$

$$\text{for } k = \{\text{Na}^+, \text{K}^+, \text{Ca}^{2+} \text{ and } \text{Cl}^-\}.$$

In this system, the variables are the electric potential, $\phi$ (in mV), and the ion concentrations, $c_k$ (in mM) for $k = \{\text{Na}^+, \text{K}^+, \text{Ca}^{2+}, \text{Cl}^-\}$. Furthermore, we have used the following definitions,

$$\varepsilon = \varepsilon_r \varepsilon_0, \tag{12.3}$$

$$\rho = \rho_0 + F \sum_k z_k c_k, \tag{12.4}$$

$$\beta_k = \frac{z_k e}{k_B T}. \tag{12.5}$$

The parameters $F, \varepsilon_0, \varepsilon_r, \rho_0, D_k, e, k_B, T$, and $z_k$ for $k = \{\text{Na}^+, \text{K}^+, \text{Ca}^{2+} \text{ and } \text{Cl}^-\}$ are defined in Table 12.1.

**Table 12.1** Parameter values used in the PNP simulations, taken from [10].

| Parameter | Description | Value |
|---|---|---|
| $F$ | Faraday's constant | 96485.3365 C/mol |
| $\varepsilon_0$ | Vacuum permittivity | 8854 fF/m |
| $\varepsilon_1$ | Relative permittivity, $\varepsilon_r$, in $\Omega_i$ and $\Omega_e$ | 80 |
| $\varepsilon_m$ | Relative permittivity, $\varepsilon_r$, in $\Omega_m, \Omega_c$ | 2 |
| $D_{\text{Na}^+}$ | Diffusion coefficient for Na$^+$ in $\Omega_i$ and $\Omega_e$ | $1.33 \cdot 10^6$ nm$^2$/ms |
| $D_{\text{K}^+}$ | Diffusion coefficient for K$^+$ in $\Omega_i$ and $\Omega_e$ | $1.96 \cdot 10^6$ nm$^2$/ms |
| $D_{\text{Ca}^{2+}}$ | Diffusion coefficient for Ca$^{2+}$ in $\Omega_i$ and $\Omega_e$ | $0.71 \cdot 10^6$ nm$^2$/ms |
| $D_{\text{Cl}^-}$ | Diffusion coefficient for Cl$^-$ in $\Omega_i$ and $\Omega_e$ | $2.03 \cdot 10^6$ nm$^2$/ms |
| $z_{\text{Na}^+}$ | Valence of Na$^+$ | 1 |
| $z_{\text{K}^+}$ | Valence of K$^+$ | 1 |
| $z_{\text{Ca}^{2+}}$ | Valence of Ca$^{2+}$ | 2 |
| $z_{\text{Cl}^-}$ | Valence of Cl$^-$ | $-1$ |
| $e$ | Elementary charge | $1.60217662 \cdot 10^{-19}$ C |
| $k_B$ | Boltzmann constant | $1.380649 \cdot 10^{-20}$ mJ/K |
| $T$ | Temperature | 310 K |
| $\Delta x, \Delta y$ | Spatial discretization parameter | 0.5 nm |
| $\Delta t$ | Numerical time step | 0.02 ns |

Again, we will solve this system by applying operator splitting and replacing derivatives by differences. We assume that the solution is known at time $t_n = n\Delta t$. The

**Table 12.2** Initial conditions for the ion concentrations in the intracellular space, $\Omega_i$, and the extracellular space, $\Omega_e$. In the membrane, $\Omega_m$, all ion concentrations are set to zero. Furthermore, in the K$^+$ channel embedded in the membrane, the concentration of K$^+$ ions varies linearly from the intracellular to the extracellular concentration, whereas the remaining ion concentrations are set to zero. To make the entire domain electroneurtral at $t = 0$, $\rho_0$ is set up so that $\rho = 0$ in the channel for the initial conditions (see (12.4)).

| Ion | Intracellular | Extracellular |
|-----|---------------|---------------|
| Na$^+$ | 12 mM | 100 mM |
| K$^+$ | 125 mM | 5 mM |
| Ca$^{2+}$ | 0.0001 mM | 1.4 mM |
| Cl$^-$ | 137.0002 mM | 107.8 mM |

first step in the algorithm is to compute the electrical potential at time $t_{n+1} = (n+1)\Delta t$. This step can be written as

$$\nabla_h \cdot (\varepsilon \nabla_h \phi^{n+1}) = -\rho^n, \tag{12.6}$$

where $\nabla_h$ denotes a finite difference approximation of the gradient. Note that $\rho^n$ is taken from the previous time step so only the electrical potential is unknown in this step. In the first time step, we use the initial conditions to compute $\rho$ given by (12.4).

When $\phi^{n+1}$ has been computed, we can compute $\nabla_h \phi^{n+1}$ and use it to solve the concentration equations by the following scheme,

$$\frac{c_k^{n+1} - c_k^n}{\Delta t} = \nabla_h \cdot D_k \nabla_h c_k^{n+1} + \nabla_h \cdot \left( D_k \beta_k c_k^{n+1} \nabla_h \phi^{n+1} \right), \tag{12.7}$$

for $k = \{$Na$^+$, K$^+$, Ca$^{2+}$,Cl$^-\}$. Writing the complete finite difference schemes for these equations is a bit messy, but the interested reader can consult the online Matlab code, or the supplementary information of [10].

## 12.1.1 Numerical Simulation of the Resting State

We will use the scheme given by (12.6) and (12.7) to compute the resting state close to the cell membrane. In the models introduced in the previous chapters, we have taken the concentrations to be constants in space; i.e., we have assumed that the concentrations can vary in time across the cell membrane, but not in space in the intra- or extracellular spaces. This is often an accurate approximation, but we will see that significant gradients exist very close to the cell membrane. Furthermore, in the EMI model, the cell membrane is assumed to be infinitely thin, and in the bidomain and monodomain models, the cell membrane is assumed to be everywhere! In the PNP model simulation, the cell membrane is explicitly represented in the model

**Fig. 12.1** Illustration of
the PNP model domain,
consisting of an intracellular
domain, $\Omega_i$, an extracellular
domain, $\Omega_e$, a membrane
domain, $\Omega_m$, and a K$^+$
channel domain, $\Omega_c$. In
the simulation reported in
Fig. 12.2, we use the domain
size $L_i = L_e = L_y = 50$ nm,
$L_m = 5$ nm, $w_c = 5$ nm.

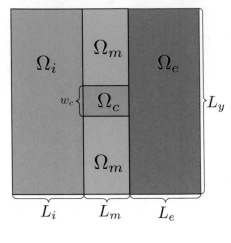

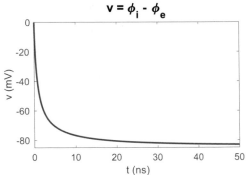

**Fig. 12.2** The membrane
potential, $v = \phi_i - \phi_e$, as a
function of time in the PNP
simulation.

(5 nm wide), but still the ion channels integrated in the membrane are modeled in
an oversimplified manner.

We solve the PNP system in two spatial dimensions, see Fig. 12.1. Note that the
computational domain consists of an intracellular domain, $\Omega_i$, a cell membrane, $\Omega_m$,
an extracellular space, $\Omega_e$, and a K$^+$ channel, $\Omega_c$. The initial conditions are given in
Table 12.2, and the solutions are presented in Figs. 12.2–12.4.

In Fig. 12.2, we have plotted the membrane potential ($v = \phi_i - \phi_e$) as a function
of time during the PNP simulation, and we see that the value starts at 0 and gradually
approaches a typical resting potential value of about −80 mV. In Fig. 12.3 and
Fig. 12.4, we show how the PNP model solution varies in space at the end of the
simulation (at $t = 50$ ns) in the part of the domain that is located in the 5 nm closest
to the membrane. The upper panel focuses on the intracellular side and the lower
panel focuses on the extracellular side of the membrane. In Fig. 12.3, we plot the
full 2D solution, and in Fig. 12.4, we plot the solution along lines in the $x$-direction
at $y = 0$ nm and $y = 25$ nm. We observe that a boundary layer is formed close to the
membrane with slightly different values of the ion concentrations and potential than
in the bulk intracellular and extracellular spaces.

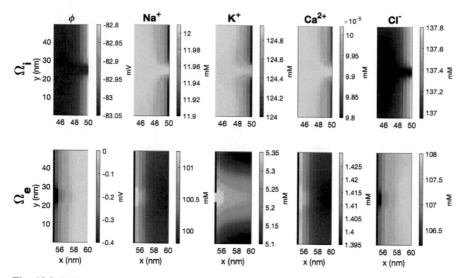

**Fig. 12.3** Solutions at the end of the simulation ($t = 50$ ns) of the PNP model simulation in the 5 nm closest to the membrane on the intracellular side (upper panel) and the extracellular side (lower panel).

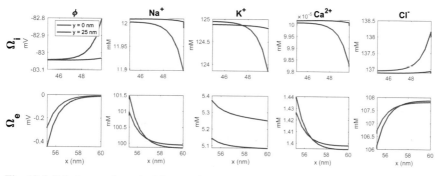

**Fig. 12.4** Solutions at the end of the simulation ($t = 50$ ns) of the PNP model simulation in the 5 nm closest to the membrane on the intracellular side (upper panel) and the extracellular side (lower panel). We show the solutions along lines in the $x$-direction for $y = 0$ nm and $y = 25$ nm.

## 12.2  Comments and Further Reading

1. The PNP equations modeling the electrical potential and ionic concentrations close to biological membranes have been studied by several authors; see, e.g., [4, 5, 10, 11, 14]. But the PNP equations are also used to model lithium ion batteries, see, e.g., [12, 17].
2. The model, methods and setup in this chapter was motived by the paper [10].
3. A simplified version of the PNP equations are referred to as the KNP (Kirchhoff-Nernst-Planck) equations. In these equations, electroneutrality is

assumed everywhere, meaning that $\rho \equiv 0$ in the entire domain. That is, the charges sum to zero everywhere. Numerical approximations of the KNP equations can be found in, e.g., [2, 3, 13, 15].

4. Above, we mentioned that one single computational block ($\Delta x^3 = (0.25 \text{ mm})^3$) for a standard mesh used to solve the bidomain model is large enough to cover almost 1000 cardiomyocytes. In the PNP model we use the resolution $\Delta x = 0.5$ nm and thus the volume of one block is $0.125$ nm$^3$. The volume of a sodium atom is $\approx 0.0244$ nm$^3$ so one computational block covers about five sodium atoms. The next scale, following bidomain/EMI/PNP, is therefore simulation based on representation of individual atoms. If a cell with a volume of 16 pL is represented by a uniform mesh at atomic (sodium atom) resolution, it will require about $6.5 \times 10^{14}$ blocks, which is a lot! A reasonably well-equipped PC today has 16 GB memory and can therefore work with a vector (in Matlab) of $\sim 2 \times 10^9$ real numbers. Thus, about 325,000 of these PCs are needed to store one real number per atom in a cardiomyocyte of 16 pL. So, it will probably take some time before atomic scale simulations can be used to simulate whole cells.

# References

[1] Clayton R, Panfilov A (2008) A guide to modelling cardiac electrical activity in anatomically detailed ventricles. Progress in Biophysics and Molecular Biology 96(1-3):19–43

[2] Ellingsrud AJ, Solbrå A, Einevoll GT, Halnes G, Rognes ME (2020) Finite element simulation of ionic electrodiffusion in cellular geometries. Frontiers in Neuroinformatics 14:11

[3] Ellingsrud AJ, Daversin-Catty C, Rognes ME (2021) A cell-based model for ionic electrodiffusion in excitable tissue. In: Modeling Excitable Tissue, Springer, pp 14–27

[4] Gardner CL, Jones JR (2011) Electrodiffusion model simulation of the potassium channel. Journal of Theoretical Biology 291:10–13

[5] Gardner CL, Nonner W, Eisenberg RS (2004) Electrodiffusion model simulation of ionic channels: 1d simulations. Journal of Computational Electronics 3(1):25–31

[6] Hille B (2001) Ion channels of excitable membranes, vol 507. Sinauer Sunderland

[7] Jæger KH, Tveito A (2021) Derivation of a cell-based mathematical model of excitable cells. In: Modeling Excitable Tissue, Springer, pp 1–13

[8] Jæger KH, Tveito A (2022) Deriving the bidomain model of cardiac electrophysiology from a cell-based model; properties and comparisons. Frontiers in Physiology 12:811029

[9] Jæger KH, Edwards AG, Giles WR, Tveito A (2022) Arrhythmogenic influence of mutations in a myocyte-based computational model of the pulmonary vein sleeve. Scientific Reports 12(1):1–18

[10] Jæger KH, Ivanovic E, Kucera JP, Tveito A (2023) Nano-scale solution of the Poisson-Nernst-Planck (PNP) equations in a fraction of two neighboring cells reveals the magnitude of intercellular electrochemical waves. PLoS Computational Biology 19(2):e1010895

[11] Jasielec JJ (2021) Electrodiffusion phenomena in neuroscience and the nernst–planck–poisson equations. Electrochem 2(2):197–215

[12] Landstorfer M, Jacob T (2013) Mathematical modeling of intercalation batteries at the cell level and beyond. Chemical Society Reviews 42(8):3234–3252

[13] Mori Y, Peskin C (2009) A numerical method for cellular electrophysiology based on the electrodiffusion equations with internal boundary conditions at membranes. Communications in Applied Mathematics and Computational Science 4(1):85–134

[14] Pods J (2017) A comparison of computational models for the extracellular potential of neurons. Journal of Integrative Neuroscience 16(1):19–32

[15] Solbrå A, Bergersen AW, van den Brink J, Malthe-Sørenssen A, Einevoll GT, Halnes G (2018) A Kirchhoff-Nernst-Planck framework for modeling large scale extracellular electrodiffusion surrounding morphologically detailed neurons. PLoS Computational Biology 14(10):e1006510

[16] Xie F, Qu Z, Yang J, Baher A, Weiss JN, Garfinkel A, et al. (2004) A simulation study of the effects of cardiac anatomy in ventricular fibrillation. The Journal of Clinical Investigation 113(5):686–693

[17] Zhao Y, Stein P, Bai Y, Al-Siraj M, Yang Y, Xu BX (2019) A review on modeling of electro-chemo-mechanics in lithium-ion batteries. Journal of Power Sources 413:259–283

# Index

Printed in the United States
by Baker & Taylor Publisher Services